REINFORCEMENT AND SYSTEMIC MACHINE LEARNING FOR DECISION MAKING

Reinforcement and Systemic Machine Learning for Decision Making

Parag Kulkarni

IEEE PRESS

A JOHN WILEY & SONS, INC., PUBLICATION

Library of Congress Cataloging-in-Publication Data:

Kulkarni, Parag.
    Reinforcement and systemic machine learning for decision making / Parag
Kulkarni.
        p. cm. – (IEEE series on systems science and engineering ; 1)
        ISBN 978-0-470-91999-6
    1. Reinforcement learning.   2. Machine learning.   3. Decision Making. I.
Title.
    Q325.6.K85 2012
    006.3′1–dc23
                                                            2011043300

Printed in the United States of America

10 9 8 7 6 5 4 3 2 1

*Dedicated to the late D.B. Joshi and the late Savitri Joshi,
who inspired me to think differently*

CONTENTS

There has been movement for years to make machines intelligent. This movement began long ago, even long before the computer era. Event-based intelligence in those days was incorporated in appliances or the ensemble of appliances. This intelligence was very much guided, and human intervention was mandatory. Even feedback control systems are a rudimentary form of intelligent system. Later adaptive control systems and hybrid control systems added flair of intelligence in these systems. This movement has received more attention with the advent of computers. Simple event-based learning with computers became a part of many intelligent systems very quickly. The expectation from intelligent systems kept on increasing. This led to one of the very well-received paradigms of learning, which is pattern-based learning. This allowed the systems to exhibit intelligence in many practical scenarios. It included patterns of weather, patterns of occupancy, and different patterns where patterns could help to make decisions. This paradigm evolved into a paradigm of behavioral pattern-based learning. This was more a behavioral pattern than a simple pattern of a particular measurement parameter. Behavioral patterns attempted to give a better picture and insight. This helped to learn and make decisions in case of networks and business scenarios. This took the intelligent systems to the next level. Learning is a manifestation of intelligence. Making machines to learn is a major part of the movement to make machines intelligent.

The complexities in decision scenarios and making machines to learn in complex scenarios raised many questions on the intelligence of a machine. Learning in isolation is never complete. Human beings learn in groups, develop colonies, and interact to build intelligence. The collective and cooperative learning of humans allows them to achieve supremacy. Furthermore, humans learn in association with the environment. They interact with the environment and receive feedback in the form of a reward or penalty. Their learning in association gives them power for exploration-based learning. Exploitation of already learned facts and exploration with reference to actions takes place. The paradigm of reinforcement learning added a new dimension to learning and could cover many new aspects of learning required for dynamic scenarios.

As mentioned by Rutherford D. Roger: "We are drowning in information and starving for knowledge." More and more information becomes available for our disposal. This information is in heterogeneous forms. There are many information sources and numerous learning opportunities. The practical assumptions while learning can make learning restrictive. Actually there are relationships among different parts of the system, and one of the basic principles of system thinking states is that the cause and effect are separated in time and space. The impact of the decision or any action can be felt beyond visible boundaries. Failing to consider this systemic aspect and relationship will lead to many limitations while learning, and hence the traditional learning paradigms suffer in highly dynamic and complex real-life problems. The holistic view and understanding of the interdependencies and intradependencies can help us to learn many new aspects and understand, analyze, and interpret the information in a more realistic way. The aspect of learning based on available information, building new information, mapping it to knowledge, and understanding different perspectives while learning can really help to make learning more effective. Learning is not just getting more data and arranging that data. It is not even building more information. Basically, the purpose of learning is to empower individuals to make better decisions and to improve their ability to create value. In machine learning, there is a need to expand the ability of machines with reference to different information sources and learning opportunities. In machine learning, it is also about empowering machines to make better decisions and improving their ability to create value.

This book is an attempt to put forth a new paradigm of systemic machine learning and research opportunities in machine learning with reference to different aspects of machine learning. The book tries to build the foundation for systemic machine learning with elaborate case studies. Machine learning and artificial intelligence are interdisciplinary in nature. Right from statistics, mathematics, psychology, and computer engineering, many researchers contributed to this field to make it rich and achieve better results. Based on these numerous contributions and our research in machine learning field, this book tries to explore the concept of systemic machine learning. Systemic machine learning is holistic, multiperspective, incremental, and systemic. While learning we can learn different things from the same data sets, we can also learn from already learned facts, and there can be number of representations of knowledge. This book is an attempt to build a framework to make the best use of all information sources and build knowledge with reference to the complete system.

In many cases, the problem is not static. It changes with time and depends on environment, and the solution even depends on the decision context. Context may not be just limited to a few parameters, but the overall information about a problem builds the context. A general-purpose system without context may not be able to handle context-specific decision. This book discusses different facets of learning as well as the need of a new paradigm with reference to complex decision problems. The book can be used as a reference book for specialized research and can help readers and researchers to appreciate new paradigms of machine learning.

This book is organized as depicted in the following figure:

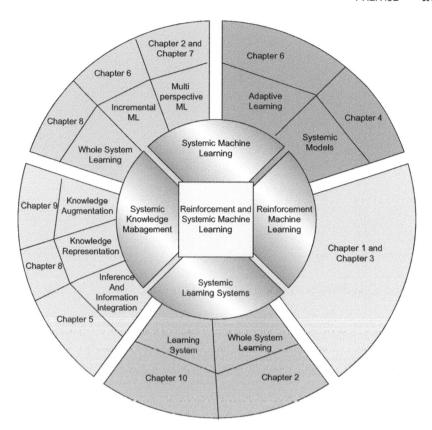

Chapter 1 introduces concepts of systemic and reinforcement machine learning. It builds a platform for the paradigm of systemic machine learning while highlighting the need of the same. Chapter 2 throws more light on the fundamentals of systemic machine learning, whole system learning, and multiperspective learning. Chapter 3 is about reinforcement learning while Chapter 4 deals with systemic machine learning and model building. The important aspects of decision making such as inference are covered in Chapter 5. Chapter 6 discusses adaptive machine learning and various aspects of adaptive machine learning. Chapter 7 discusses the paradigm of multiperspective machine learning and whole system learning. Chapter 8 addresses the need for incremental machine learning. Chapters 8 and 9 deal with knowledge representation and knowledge augmentation. Chapter 10 discusses the building learning system.

This book tries to include different facets of learning while introducing a new paradigm of machine learning. It deals with building knowledge through machine learning. This book is for those individuals who are planning to contribute to make a machine more intelligent by making it learn through new experiments, are ready to try new ways, and are open for a new paradigm for the same.

PARAG KULKARNI

ACKNOWLEDGMENTS

For the past two decades I have been working with various decision-making and AI-based IT product companies. During this period I worked on different Machine Learning algorithms and applied them for different applications. This work made me realize the need for a new paradigm for machine learning and the need for change in thinking. This built the foundation for this book and started the thought process for systemic machine learning. I am thankful to different organizations I worked with, including Siemens and IDeaS, and to my colleagues in those organizations. I would also like to acknowledge the support of my friends and coworkers.

I would like to thank my Ph.D. and M.Tech. students—Prachi, Yashodhara, Vinod, Sunita, Pramod, Nitin, Deepak, Preeti, Anagha, Shankar, Shweta, Basawraj, Shashikanth, and others—for their direct and indirect contribution that came through technical brainstorming. They are always ready to work on new ideas and contributed through collective learning. Special thanks to Prachi for her help in drawing diagrams and formatting the text.

I am thankful to Prof. Chande, the late Prof. Ramani, Dr. Sinha, Dr. Bhanu Prasad, Prof. Warnekar, and Prof. Navetia for useful comments and reviews. I am also thankful to Institutes such as COEP, PICT, GHRIET, PCCOE, DYP COE, IIM, Masaryk University, and so on, for allowing me to interact and present my thoughts in front of students. I am also thankful to IASTED, IACSIT, and IEEE for giving me the platform to present my research through various technical conferences. I am also thankful to reviewers of my research papers.

I am thankful to my mentor, teacher, and grandfather, the late D.B. Joshi, for motivating me to think differently. I also would like to take the opportunity to thank my mother. Most importantly I would like to thank my wife Mrudula and son Hrishikesh for their support, motivation, and help.

I am also thankful to IEEE/Wiley and the editorial team of IEEE/Wiley for their support and helping me to present my research, thoughts, and experiments in the form of a book.

PARAG KULKARNI

About the Author

Parag Kulkarni, Ph.D. D.Sc., is CEO and Chief Scientist at EKLaT Research, Pune. He has more than two decades of experience in knowledge management, e-business, intelligent systems and machine learning consultation, research and product building. An alumnus of IIT Kharagpur and IIM Kolkata, Dr. Kulkarni has been a visiting professor at IIM Indore, visiting researcher at Masaryk University Czech Republic, and Adjunct Professor at the College of Engineering, Pune. He has headed companies, research labs, and groups at various IT companies including IDeaS, Siemens Information Systems Ltd., and Capilson, Pune, and ReasonEdge, Singapore. He has led many start-up companies to success through strategic innovation and research. The UGSM Monarch Business School, Switzerland, has conferred higher doctorate D.Sc. on Dr. Kulkarni. He is a coinventor of three patents and has coauthored more than 100 research papers and several books.

CHAPTER 1

Introduction to Reinforcement and Systemic Machine Learning

1.1 INTRODUCTION

The expectations from intelligent systems are increasing day by day. What an intelligent system was supposed to do a decade ago is now expected from an ordinary system. Whether it is a washing machine or a health care system, we expect it to be more and more intelligent and demonstrate that behavior while solving complex as well as day-to-day problems. The applications are not limited to a particular domain and are literally distributed across all domains. Hence domain-specific intelligence is fine but the user has become demanding, and a true intelligent and problem-solving system irrespective of domains has become a necessary goal. We want the systems to drive cars, play games, train players, retrieve information, and help even in complex medical diagnosis. All these applications are beyond the scope of isolated systems and traditional preprogrammed learning. These activities need dynamic intelligence. Dynamic intelligence can be exhibited through learning not only based on available knowledge but also based on the exploration of knowledge through interactions with the environment. The use of existing knowledge, learning based on dynamic facts, and acting in the best way in complex scenarios are some of the expected features of intelligent systems.

The learning has many facets. Right from simple memorization of facts to complex inference are some examples of learning. But at any point of time, learning is a holistic activity and takes place around the objective of better decision-making. Learning results from data storing, sorting, mapping, and classification. Still one of the most important aspects of intelligence is learning. In most of the cases we expect learning to be a more goal-centric activity. Learning results from an inputs from an experienced person, one's own experience, and inference based on experiences or past learning. So there are three ways of learning:

- Learning based on expert inputs (supervised learning)

Reinforcement and Systemic Machine Learning for Decision Making, First Edition. Parag Kulkarni.
© 2012 by the Institute of Electrical and Electronics Engineers, Inc.
Published 2012 by John Wiley & Sons, Inc.

- Learning based on own experience
- Learning based on already learned facts

In this chapter, we will discuss the basics of reinforcement learning and its history. We will also look closely at the need of reinforcement learning. This chapter will discuss limitations of reinforcement learning and the concept of systemic learning. The systemic machine-learning paradigm is discussed along with various concepts and techniques. The chapter also covers an introduction to traditional learning methods. The relationship among different learning methods with reference to systemic machine learning is elaborated in this chapter. The chapter builds the background for systemic machine learning.

1.2 SUPERVISED, UNSUPERVISED, AND SEMISUPERVISED MACHINE LEARNING

Learning that takes place based on a class of examples is referred to as *supervised learning*. It is learning based on labeled data. In short, while learning, the system has knowledge of a set of labeled data. This is one of the most common and frequently used learning methods. Let us begin by considering the simplest machine-learning task: *supervised learning* for classification. Let us take an example of classification of documents. In this particular case a learner learns based on the available documents and their classes. This is also referred to as labeled data. The program that can map the input documents to appropriate classes is called a *classifier*, because it assigns a class (i.e., document type) to an object (i.e., a document). The task of supervised learning is to construct a classifier given a set of classified training examples. A typical classification is depicted in Figure 1.1.

Figure 1.1 represents a hyperplane that has been generated after learning, separating two classes—class A and class B in different parts. Each input point presents input–output instance from sample space. In case of document classification, these points are documents. Learning computes a separating line or hyperplane among documents. An unknown document type will be decided by its position with respect to a separator.

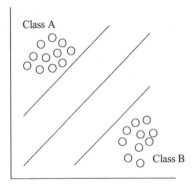

Figure 1.1   Supervised learning.

There are a number of challenges in supervised classification such as generalization, selection of right data for learning, and dealing with variations. *Labeled examples* are used for training in case of supervised learning. The set of labeled examples provided to the learning algorithm is called the *training set*.

The classifier and of course the decision-making engine should minimize false positives and false negatives. Here false positives stand for the result yes—that is, classified in a particular group wrongly. False negative is the case where it should have been accepted as a class but got rejected. For example, apples not classified as apples is false negative, while an orange or some other fruit classified as an apple is false positive in the apple class. Another example of it is when guilty but not convicted is false positive, while innocent but convicted or declared innocent is false negative. Typically, wrongly classified are more harmful than unclassified elements.

If a classifier knew that the data consisted of sets or batches, it could achieve higher accuracy by trying to identify the boundary between two adjacent sets. It is true in the case of sets of documents to be separated from one another. Though it depends on the scenario, typically false negatives are more costly than false positives, so we might want the learning algorithm to prefer classifiers that make fewer false negative errors, even if they make more false positives as a result. This is so because false negative generally takes away the identity of the objects or elements that are classified correctly. It is believed that the false positive can be corrected in next pass, but there is no such scope for false negative.

Supervised learning is not just about classification, but it is the overall process that with guidelines maps to the most appropriate decision.

Unsupervised learning refers to learning from unlabeled data. It is based more on similarity and differences than on anything else. In this type of learning, all similar items are clustered together in a particular class where the label of a class is not known.

It is not possible to learn in a supervised way in the absence of properly labeled data. In these scenarios there is need to learn in an unsupervised way. Here the learning is based more on similarities and differences that are visible. These differences and similarities are mathematically represented in unsupervised learning.

Given a large collection of objects, we often want to be able to understand these objects and visualize their relationships. For an example based on similarities, a kid can separate birds from other animals. It may use some property or similarity while separating, such as the birds have wings. The criterion in initial stages is the most visible aspects of those objects. Linnaeus devoted much of his life to arranging living organisms into a hierarchy of classes, with the goal of arranging similar organisms together at all levels of the hierarchy. Many unsupervised learning algorithms create similar hierarchical arrangements based on similarity-based mappings. The task of *hierarchical clustering* is to arrange a set of objects into a hierarchy such that similar objects are grouped together. *Nonhierarchical clustering* seeks to partition the data into some number of disjoint clusters. The process of clustering is depicted in Figure 1.2. A learner is fed with a set of scattered points, and it generates two clusters with representative centroids after learning. Clusters show that points with similar properties and closeness are grouped together.

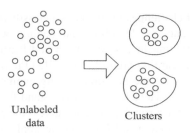

Unlabeled data   Clusters

Figure 1.2   Unsupervised learning.

In practical scenarios there is always need to learn from both labeled and unlabeled data. Even while learning in an unsupervised way, there is the need to make the best use of labeled data available. This is referred to as *semisupervised learning*. Semisupervised learning is making the best use of two paradigms of learning—that is, learning based on similarity and learning based on inputs from a teacher. Semisupervised learning tries to get the best of both the worlds.

1.3   TRADITIONAL LEARNING METHODS AND HISTORY OF MACHINE LEARNING

Learning is not just knowledge acquisition but rather a combination of knowledge acquisition, knowledge augmentation, and knowledge management. Furthermore, intelligent inference is essential for proper learning. Knowledge deals with significance of information and learning deals with building knowledge. How can a machine can be made to learn? This research question has been posed for more than six decades by researchers. The outcome of this research has built a platform for this chapter. Learning involves every activity. One such example, is the following: While going to the office yesterday, Ram found road repair work in progress on route one, so he followed route two today. It might be possible that route two is worse. Then he may go back to route one or might try route three. Route one is in bad shape due to repair work is knowledge built, and based on that knowledge he has taken action: following route 2, that is, exploration. The complexity of learning increases as the number of parameters and time dimensions start playing a role in decision making.

Ram found that road repair work is in progress on route one.

He hears an announcement that in case of rain, route two will be closed.

He needs to visit a shop X while going to office.

He is running out of petrol.

These new parameters make his decision much more complex as compared to scenario 1 and scenario 2 discussed above.

In this chapter, we will discuss various learning methods along with examples. The data and information used for learning are very important. The data cannot be used as is for learning. It may contain outliers and information about features that may not be relevant with respect to the problem one is trying to solve. The approaches for the selection of data for learning vary with the problems. In some cases the most frequent patterns are used for learning. Even in some cases, outliers are also used for learning. There can be learning based on exceptions. The learning can take place based on similarities as well as differences. The positive as well as negative examples help in effective learning. Various models are built for learning with the objective of exploiting the knowledge.

Learning is a continuous process. The new scenarios are observed and new situations arise—those need to be used for learning. Learning from observation needs to construct meaningful classification of observed objects and situation. Methods of measuring similarity and proximity are employed for this purpose. Learning from observations is the most commonly used method by human beings. While making decisions we may come across the scenarios and objects that we have not used or came across during a learning phase. The inference allows us to handle these scenarios. Furthermore, we need to learn in different and new scenarios and hence even while making decisions the learning continues.

There are three fundamental continuously active human-like learning mechanisms:

1. *Perceptual Learning:* Learning of new objects, categories, and relations. It is more like constantly seeking to improve and grow. It is similar to the learning professionals use.
2. *Episodic Learning:* It is based on events and information about the event, like what, where, and when. It is the learning or the change in the behavior that occurs due to an event.
3. *Procedural Learning:* Learning based on actions and action sequences to accomplish a task. Implementation of this human cognition can impart intelligence to a machine. Hence, a unified methodology around intelligent behavior is the need of time that will allow machines to learn and behave or respond intelligently in dynamic scenarios.

Traditional machine-learning approaches are susceptible to dynamic continual changes in the environment. However, perceptual learning in human does not have such restrictions. Learning in humans is selectively incremental, so it does not need a large training set and is simultaneously not biased by already learned but outdated facts. Learning and knowledge extraction in human beings is dynamic, and a human brain adapts to changes occurring in the environment continuously.

Interestingly, psychologists have played a major role in the development of machine-learning techniques. It has been a movement taken by computer researchers and psychologists together to make machines intelligent for more than six decades. The application areas are growing, and research done in the last six decades made us believe that it is one of the most interesting areas to make machines learn.

Machine learning is the study of methods for programming computers to learn. It is about making machines to behave intelligently and learn from experiences like human beings. In some tasks the human expert may not be required; this may include automated manufacturing or repetitive tasks with very few dynamic situations but demanding very high level of precision. A machine-learning system can study recorded data and subsequent machine failures and learn prediction rules. Second, there are problems where human experts exist and are required, but the knowledge is present in a tacit form. Speech recognition and language understanding come under this category. Virtually all humans exhibit expert-level abilities on these tasks, but the exact method and steps to perform these tasks are not known. A set of inputs and outputs with mapping is provided in this case, and thus machine-learning algorithms can learn to map the inputs to the outputs.

Third, there are problems where phenomena are changing rapidly. In real life there are many dynamic scenarios. Here the situations and parameters are changing dynamically. These behaviors change frequently, so that even if a programmer could construct a good predictive computer program, it would need to be rewritten frequently. A learning program can relieve the programmer of this burden by constantly modifying and tuning a set of learned prediction rules.

Fourth, there are applications that need to be customized for each computer user separately. A machine-learning system can learn the customer-specific requirements and tune the parameters accordingly to get a customized version for a specific customer.

Machine learning addresses many of the research questions with the aid of statistics, data mining, and psychology. Machine learning is much more than just data mining and statistics. Machine learning (ML) as it stands today is the use of data mining and statistics for inferencing to make decisions or build knowledge to enable better decision making. Statistics is more about understanding data and the pattern between them. Data mining seeks the relevant data based on patterns for decision making and analysis. Psychological studies of human learning aspire to understand the mechanisms underlying the various learning behaviors exhibited by people. At the end of the day, we want machine learning to empower machines with the learning abilities that are demonstrated by humans in complex scenarios. The psychological studies of human nature and the intelligence also contribute to different methods of machine learning. This includes concept learning, skill acquisition, strategy change, analytical inferences, and bias based on scenarios.

Machine learning is primarily concerned with the timely response, accuracy, and effectiveness of the resulting computer system. It many times does not take into account other aspects such as learning abilities and responding to dynamic situations, which are equally important. A machine-learning approach focuses on many complex applications such as building an accurate face recognition and authentication system. Statisticians, psychologists, and computer scientists may work together on this front. A data mining approach might look for patterns and variations in image data.

One of the major aspects of learning is the selection of learning data. All the information available for learning cannot be used as it is. It may contain a lot of data that may not be relevant or captured from a completely different perspective. Every bit of data cannot be used with the same importance and priority. The prioritization of the data is done based on scenarios, system significance, and relevance. The determination of relevance of these data is one of the most difficult parts of the process.

There are a number of challenges in making machines learn and making suitable decisions at the right time. The challenges start from the availability of limited learning data, unknown perspectives, and defining the decision problems. Let us take a simple example where a machine is expected to prescribe the right medicine to a patient. The learning set may include samples of patients, their histories, their test reports, and the symptoms reported by them. Furthermore, the data for learning may also include other information such as family history, habits, and so on. In case of a new patient, there is the need to infer based on available limited information because the manifestation of the same disease may be different in his case. Some key information might be missing, and hence decision making may become even more difficult.

When we look at the way a human being learns, we find many interesting aspects. Generally the learning takes place with understanding. It is facilitated when new and existing knowledge is structured around the major concepts and principles of the discipline. During the learning, either some principles are already there or developed in the process work as a guideline for learning. The learning also needs prior knowledge. Learners use what they already know to construct new understandings. This is more like building knowledge. Furthermore, there are different perspectives and metacognition. Learning is facilitated through the use of metacognitive strategies that identify, monitor, and regulate cognitive processes.

1.4   WHAT IS MACHINE LEARNING?

A general concept of machine learning is depicted in Figure 1.3. Machine learning studies computer algorithms for learning. We might, for instance, be interested in learning to complete a task, or to make accurate predictions, reactions in certain situations, or to behave intelligently. The learning that is being done is always based on some sort of observations or data, such as examples (the most common case in this course), direct experience, or instruction. So in general, machine learning is about learning to do better in the future based on what was experienced in the past. It is making a machine to learn from available information, experience, and knowledge building.

In the context of the present research, machine learning is the development of programs that allow us to analyze data from the various sources, select relevant data,

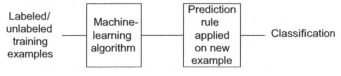

Figure 1.3   Machine learning and classification.

and use those data to predict the behavior of the system in another similar and if possible different scenario. Machine learning also classifies objects and behaviors to finally impart the decisions for new input scenarios. The interesting part is that more learning and intelligence is required to deal with uncertain situations.

1.5 MACHINE-LEARNING PROBLEM

It can be easily concluded that all the problems that need intelligence to solve come under the category of machine-learning problems. Typical problems are character recognition, face authentication, document classification, spam filtering, speech recognition, fraud detection, weather forecasting, and occupancy forecasting. Interestingly, many problems that are more complex and involve decision making can be considered as machine-learning problems as well. These problems typically involve learning from experiences and data, and search for the solutions in known as well as unknown search spaces. It may involve the classification of objects, problems, and mapping them to solutions or decisions. Even classification of any type of objects or events is also a machine-learning problem.

1.5.1 Goals of Learning

The primary goal of learning/machine learning is producing some learning algorithm with practical value. In the literature and research, most of the time machine learning is referred to from the perspective of applications and it is more bound by methods. The goals of ML are described as development and enhancement of computer algorithms and models to meet the decision-making requirements in practical scenarios. Interestingly, it did achieve the set goal in many applications. Right from washing machines and microwave ovens to the automated landing of aircraft, machine learning is playing a major role in all modern applications and appliances. The era of machine learning has introduced methods from simple data analysis and pattern matching to fuzzy logic and inferencing.

In machine learning, most of the inferencing is data driven. The sources of data are limited and many times there is difficulty in identifying the useful data. It may be possible that the source contains large piles of data and that the data contain important relationships and correlations among them. Machine learning can extract these relationships, which is an area of data mining applications. The goal of machine learning is to facilitate in building intelligent systems (IS) that can be used in solving real-life problems.

The computational power of the computing engine, the sophistication and elegance of algorithms, the amount and quality of information and values, and the efficiency and reliability of the system architecture determine the amount of intelligence. The amount of intelligence can grow through algorithm development, learning, and evolution. Intelligence is the product of natural selection, wherein more successful behavior is passed on to succeeding generations of intelligent systems and less successful behavior dies out. This intelligence helps humans and intelligent systems to learn.

In supervised learning we learn from different scenarios and expected outcomes presented as a learning material. The purpose is that if we come across a similar scenario in the future we should be in position to make appropriate or rather the best possible decisions. This is possible if we can classify a new scenario to one of the known classes or known scenarios. Enabling to classify the new scenario allows us to select an appropriate action. Learning is possible by imitation, memorization, mapping, and inference. Furthermore, induction, deduction, and example-based and observation-based learning are some other ways in which learning is possible.

Learning is driven by objective and governed by certain performance elements and their components. The clarity about the performance elements and their components, available feedback to learn the behavior of these components, and the representation of these components are necessary for learning. The agents need to learn, and components of these agents should be able to map and determine actions, extract and infer about the information related to the environment, and set goals that describe classes of states. The desired actions with reference to value or state help the system to learn. The learning takes place based on feedbacks. These feedbacks come in the form of penalties or rewards.

1.6 LEARNING PARADIGMS

An empirical learning method has three different approaches to modeling problems based on observation, data, and partial knowledge about problem domains. These approaches are more specific to problem domains. They are

1. Generative modeling
2. Discriminative modeling
3. Imitative modeling

Each of these models has their own pros and cons. They are best suited for different application areas depending on training samples and prior knowledge. Generally, learning model suitability depends on the problem scenario and available knowledge and decision complexities.

In a *generative modeling approach*, statistics provide a formal method for determining nondeterministic models by estimating joint probability over variables of problem domain. Bayesian networks are used to capture dependencies among domain variables as well as distributions among them. This partial domain knowledge combined with observations enhances the probability density function. Generative density function is then used to generate samples of different configurations of the system and to draw an inference on an unknown situation. Traditional rule-based expert systems are giving way to statistical generative approaches due to visualization of interdependencies among variables that yields better prediction than heuristic approaches. Natural language processing, speech recognition, and topic modeling among different speakers are some of the application areas of generative modeling. This probabilistic approach of learning can be used in computer vision, motion tracking, object recognition, face recognition, and so on. In a nutshell, learning with generative modeling can be applied in the domains of perception, temporal modeling, and autonomous agents. This model tries to represent and model interdependencies in order to lead to better predictions.

A *discriminative approach* models posterior probability or discriminant functions with less domain-specific or prior knowledge. This technique directly optimizes target task-related criteria. For example, a Support Vector Machine maximizes the margin of a hyperplane between two sets of variables in n dimensions. This approach can be widely used for document classification, character recognition, and other numerous areas where interdependency among problem variables does not play any role or play the minimum role in observation variables. Thus, prediction is not influenced by inherent problem structure and also by domain knowledge. This approach may not be very effective in the case of very high level of interdependencies.

The third approach is *imitative learning*. Autonomous agents, which exhibit interactive behavior, are trained through an imitative learning model. The objective of imitative leaning is to learn an agent's behavior by providing a real example of agents' interaction with the world and generalizing it. The two components of this learning model, passively perceiving real-world behavior and learning from it, are depicted in Figure 1.4. Interactive agents perceive the environment using a generative model to regenerate/synthesize virtual characters/interaction and use a discriminative approach on temporal learning to focus on the prediction task necessary for action selection. An agent tries to imitate real-world situations with intelligence so that if exact behavior is not available in a learned hypothesis, the agent can still take some action based on synthesis. Occurrence of imitative and observational learning can be used in contingence with reinforcement learning. The imitative response can be the action for the rewards in reinforcement learning.

Figure 1.4 depicts the imitative learning with reference to a demonstrator and environment. The demonstration is rather action or a series of actions from which an observer learns. Environment refers to the environment of the observer. The learning takes place based on imitation and observation of demonstration while knowledge base and environment help in inferring different facts to complete the learning. Imitative learning can be extended to imitative reinforcement learning where imitation is based on previous knowledge learning and the rewards are compared with pure imitative response.

Learning based on experience need to have input and outcome of experience to be measured. For any action there is some outcome. The outcome leads to some sort of amendment in your action. Learning can be data-based, event-based, pattern-based, and system-based. There are advantages and disadvantages of each of these paradigms of learning. Knowledge building and learning is a continuous process, and we would like systems to reuse creatively and intelligently what is learned selectively in order to achieve the goal state.

Interestingly, when a kid is learning to walk, it is using all types of learning simultaneously. It has some supervised learning in the form of parents guiding it, some unsupervised learning based on new data points it is coming across, inference for

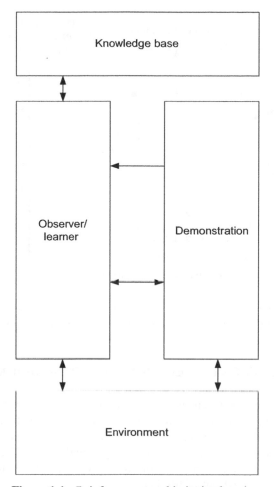

Figure 1.4   Reinforcement and imitative learning.

some similar scenarios, and feedback from environment. Learning results from labeled as well as unlabeled data, and it takes place simultaneously. In fact a kid is using all the learning methods and much more than that. A kid not only uses available knowledge and context but also infers information that cannot be derived directly from the available data. Kids use all these methods selectively, together, and based on need and appropriateness. The learning by kids results from their close interactions with environment. While making systems learn from experiences, we need to take into account all these facts. Furthermore, it is more about paradigms rather than methods used for learning. This book is about making a system intelligent with focus on reinforcement learning. Reinforcement learning tries to strike balance between exploitation and exploration. Furthermore, it takes place with interaction with environment. Rewards from environment and then cumulative value drive the overall actions. Figure 1.5 depicts the process of how a kid learns. Kids get many

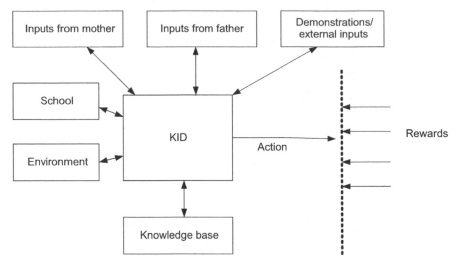

Figure 1.5 Kid learning model.

inputs from their parents, society, school, and experiences. They perform actions, and for actions they obtain rewards from these sources and environment.

1.7 MACHINE-LEARNING TECHNIQUES AND PARADIGMS

The learning paradigm kept changing over the years. The concept of intelligence changed, and even paradigm of learning and knowledge acquisition changed. Paradigm is (in the philosophy of science) a very general conception of the nature of scientific endeavor within which a given enquiry is undertaken. The learning as per Peter Senge is the acquiring of information and knowledge that can empower us to get what we would like to get out of life [1].

In machine learning if we go through the history, learning is initially assumed more as memorization and getting or reproducing one of the memorized facts that is appropriate when required. This paradigm can be called a data-centric paradigm. In fact this paradigm does exist in machine learning even today and is being used to great extent in all intelligent programs. Take the example of a simple program of retrieving the age of employees. A simple database with names and age is maintained; and when the name of any employee is given, the program can retrieve the age of the given employee. There are many such database-centric applications demonstrating data centric intelligence. But slowly the expectations from intelligent systems started increasing. As per the Turing test of intelligence, an intelligent system is one that can behave like a human, or it is difficult to make out whether a response is coming from a machine or a human.

The learning is interdisciplinary and deals with many aspects from psychology, statistics, mathematics, and neurology. Interestingly, all human behaviors could not correspond to intelligence and hence there are some areas where a computer can behave or respond in a better way. The Turing test is applicable to intelligent behavior of computers. There are some intelligent activities that humans do not do or that, machines can do in a better way than humans.

Reinforcement learning is making systems get the best of both worlds in the best possible way. But since the systemic behaviors of activities and decision making makes it necessary to understand the system behavior and components for effective decision making, traditional paradigms of machine learning may not exhibit the required intelligent behavior in complex systems. Every activity, action, and decision has some systemic impact. Furthermore, any event may result from some other event or series of events from a systemic perspective. These relationships are complex and difficult to understand. Systemic machine learning is more about exploitation, exploration from systemic perspective to build knowledge to get what we expect from the system. Learning from experience is the most important part of it. With more and more experience the behavior is expected to improve.

Two aspects of learning include learning for predictable environment behavior and learning for nonpredictable environment behavior. As we expect systems and machines to behave intelligently even in a nonpredictive environment, we need to look at learning paradigms and models from the perspective of new expectations. These expectations make it necessary to learn continuously and from various sources of information.

Representing and adapting knowledge for these systems and using them effectively is a necessary part of it. Another important aspect of learning is context: the intelligence and decision making should make effective use of context. In the absence of context, deriving the meaning of data is difficult. Further decisions may differ as per the context. Context is very systemic in nature. Context talks more about the scenario—that is, circumstances and facts surrounding the event. In absence of the facts and circumstances of related data, decision making becomes a difficult task. The context covers various aspects of environment and system such as environmental parameters, interactions with other systems and subsystems, various parameters, and so on. A doctor asks patients a number of questions. The information given by a patient along with the information with the doctor about epidemic and other recent health issues and outcome of conducted medical tests builds context for him/her. A doctor uses this context to diagnose.

The intelligence is not isolated and needs to use information from the environment for decision making as well as learning. The learning agents get feedback in the form of reward/penalty for their every action. They are supposed to learn from experience. To learn, there is a need to acquire more and more information. In real-life scenarios the agents cannot view anything and everything. There are fully observable environments and partially observable environments. Practically all environments are partially observable unless specific constraints are posed for some focused goal. The limited view limits the learning and decision-making abilities. The concept of integrating information is used very effectively in intelligent systems—the learning paradigm is confined by data-centric approaches. The context considered in the past research was more data centric and was never at a center of the activity.

1.8 WHAT IS REINFORCEMENT LEARNING?

There are tons of nonlinear and complex problems still waiting for the solutions. Ranging from automated car drivers to next level security systems. These problems look solvable—but the methods, solutions, and available information are just not enough to provide a graceful solution.

The main objective in solving a machine-learning problem is to produce intelligent programs or intelligent agents through the process of learning and adapting to changed environment. Reinforcement learning is one such machine-learning process. In this approach, learners or software agents learn from direct interaction with environment. This mimics the way human being learns. The agent can also learn even if complete model or information about environment is not available. An agent gets feedback about its actions as reward or punishment. During a learning process, these situations are mapped to actions in an environment. Reinforcement learning algorithms maximize rewards received during interactions with environment and establish the mapping of states to actions as a decision-making policy. The policy can be decided once or it can also adapt with changes in environment.

Reinforcement learning is different from *supervised learning*—the most widely used kind of learning. Supervised learning is learning from examples provided by a knowledgeable external supervisor. It is a method for training a parameterized function approximator. But it is not adequate for learning from interaction. It is more like learning from external guidance, and the guidance sits out of the environment or situation. In interactive problems it is often impractical to obtain examples of desired behavior that are both correct and representative of all the situations in which the agent has to act. In uncharted territory, where one would expect learning to be most beneficial, an agent must be able to learn from its own experience and from environment also. Thus, reinforcement learning combines the field of dynamic programming and supervised learning to generate a machine-learning system, which is very close to approaches used by human learning.

One of the challenges that arise in reinforcement learning and not in other kinds of learning is the trade-off between exploration and exploitation. To obtain a lot of reward, a reinforcement-learning agent must prefer actions that it has tried in the past and found to be effective in producing reward. But to discover such actions, it has to try actions that it has not selected before. The agent has to *exploit* what it already knows in order to obtain reward, but it also has to *explore* in order to make better action selections in the future. The dilemma is that neither exploration nor exploitation can be pursued exclusively without failing at the task. The agent must try a variety of actions *and* progressively favor those that appear to be best. On a stochastic task, each action must be tried many times to gain a reliable estimate of its expected reward. The entire issue of balancing exploration and exploitation does not arise in supervised learning, as it is usually defined. Furthermore, supervised learning never looks into exploration, and the responsibility of exploration is given to experts.

Another key feature of reinforcement learning is that it explicitly considers the *whole* problem of a goal-directed agent interacting with an uncertain environment.

This is in contrast with many approaches that consider subproblems without addressing how they might fit into a larger picture. For example, we have mentioned that much of machine-learning research is concerned with supervised learning without explicitly specifying how such ability would finally be useful. Other researchers have developed theories of planning with general goals, but without considering planning's role in real-time decision making nor considering the question of where the predictive models necessary for planning would come from. Although these approaches have yielded many useful results, their focus on isolated subproblems is a significant limitation. These limitations come from the inability to interact in real-time scenarios and the absence of active learning.

Reinforcement learning differs from the more widely studied problem of supervised learning in several ways. The most important difference is that there is no presentation of input–output pairs. Instead, after choosing an action the agent is told the immediate reward and the subsequent state, but is *not* told which action would have been in its best long-term interests. It is necessary for the agent to gather useful experience about the possible system states, actions, transitions, and rewards actively to act optimally. Another difference from supervised learning is that online performance is important; the evaluation of the system is often concurrent with learning.

Reinforcement learning takes the opposite track, starting with a complete, interactive, goal-seeking agent. All reinforcement-learning agents have explicit goals, can sense aspects of their environments, and can choose actions to influence their environments. Moreover, it is usually assumed from the beginning that the agent has to operate despite significant uncertainty about the environment it faces. When reinforcement learning involves planning, it has to address the interplay between planning and real-time action selection, as well as the question of how environmental models are acquired and improved. When reinforcement learning involves supervised learning, it does so for specific reasons that determine which capabilities are critical and which are not.

Some aspects of reinforcement learning are closely related to search and planning issues in artificial intelligence (AI), especially in the case of intelligent agents. AI search algorithms generate a satisfactory trajectory through a graph of states. The search algorithms are focused on searching a goal state based on informed or uninformed methods. The combination of informed and uninformed methods is similar to exploration and exploitation of knowledge. Planning operates in a similar manner, but typically within a construct with more complexity than a graph, in which states are represented by compositions of logical expressions instead of atomic symbols. These AI algorithms are less general than the reinforcement-learning methods, where the AI algorithms require a predefined model of state transitions and with a few exceptions assumed. These methods are typically confined by predefined models and well-defined constraints. On the other hand, reinforcement learning, at least in the form of discrete cases, assumes that the entire state space can be enumerated and stored in memory—an assumption to which conventional search algorithms are not tied.

Reinforcement learning is the problem of agents to learn from the environment by their interactions with dynamic environment. We can relate them to the learning agents. The interactions are trial and error in nature because a supervisor does not tell the agent which actions are right or wrong, unlike the case in supervised learning. There are mainly two main strategies to solve this problem. The first one is to search in behavioral space to find out the action behavior pair that performs well in the environment. The other strategy is based on statistical techniques and dynamic programming to estimate the utility of actions and chances of reaching a goal.

1.9 REINFORCEMENT FUNCTION AND ENVIRONMENT FUNCTION

As discussed above, the reinforcement learning is not just the exploitation of information based on already acquired knowledge. Rather, reinforcement learning is about balance between exploitation and exploration. Here exploitation refers to making the best use of knowledge acquired so far, while exploration refers to exploring new action, avenues, and route to build new knowledge. While exploring the action is performed, each action leads to learning through either rewards or penalties. The value function is the cumulative effect, while reward is associated with a particular atomic action. The environment needs to be modeled in changing scenarios so that it can provide the correct response that can optimize the value. The reinforcement function here is the effect of the environment, which allows the reinforcement.

Figure 1.6 depicts a typical reinforcement-learning scenario where the actions lead to rewards from environment. The purpose is to maximize expected discounted returns and also called value. The expected returns are given by

$$E\{r_{t+1} + \gamma r_{t+2} + \gamma^2 r_{t+3} + \cdots\}$$

Here discount rate is $0 \le \gamma \le 1$.

Finally the value of being in state s with reference to policy P_i is of interest to us and is calculated by

$$V^\pi(s) = E_\pi\{r_{t+1} + \gamma r_{t+2} + \gamma^2 r_{t+3} + \cdots / s_t = s\}$$

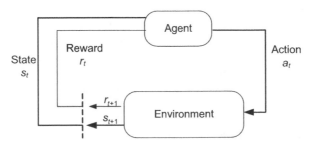

Figure 1.6 Reinforcement-learning scenario.

In short, for every action, there are environment functions and reinforcement functions. We will deal with these functions in greater detail as we proceed in this book.

1.10   NEED OF REINFORCEMENT LEARNING

Neither exploration nor exploitation alone can exhibit the intelligent learning behavior that is expected in many real-life and complex problems. A technique that makes use of both is required. While a child is learning to walk, it makes use of supervised as well as unsupervised ways of learning. Supervised here refers to inputs given to a child by parents while it may try to classify objects based on similarity and differences. Furthermore, a child explores new information by a new action and registers it. That even happens simultaneously. While kids are exploiting the knowledge, they also explore their outcomes with new actions, register learning, and build knowledge base, which is exploited in the future. In fact, exploration along with environment and learning based on the rewards or penalties is required to exhibit the intelligent behavior expected in most of the real-life scenarios.

Take an example of an intelligent automated boxing trainer. The trainer needs to exhibit more and more intelligence as it progresses and comes across a number of boxers. In addition, the trainer needs to adapt to a novice as well as expert. Furthermore, the trainer also needs to enhance his/her performance as the candidate starts exhibiting better performance. This very typical learning behavior is captured in reinforcement learning and hence it is necessary to solve many real-life problems. Learning based on data and perceived data pattern is very common. At any point of time, intelligent systems act based on percept or sequence of percepts. The percept here is the view of the intelligent system about the environment. Effective learning based on percept is required for real-time and dynamic intelligent systems. Hence machine intelligence needs learning with reference to environment, explores new paths, and exhibits intelligence in known or new scenarios. Reinforcement learning captures these requirements; hence for dynamic scenarios, reinforcement learning can be used effectively.

1.11   REINFORCEMENT LEARNING AND MACHINE INTELLIGENCE

The changing environment and environmental parameters and dynamic scenarios of many real-life problems make it difficult for a machine to arrive at solutions. If a computer could learn to solve the problems—through exploration or through trial and error—that would be of great practical value. Furthermore, there are many situations where we do not know enough about environment or problem scenarios to build an expert system, and even the correct answers are not known. The examples are car control, flight control, and so on, where there are many unknown parameters and scenarios. "Learning how to achieve the goal without knowing the exact goal till it achieves the goal" is the most complex problem intelligent systems are facing.

Reinforcement learning has one of the most important advantages for these types of problems, that is, advantage of updating.

Every moment there is a change in scenarios, and environmental parameters in the case of dynamic real-life problems. Take an example of a missile trying to hit a moving target, an automatic car driver, and business intelligent systems—in all these cases the most important aspect is learning from exploration and sensing the environment response for every progressive action. The information about the goal is revealed as we explore with help of new actions. This learning paradigm helps us in reaching a goal without the prior knowledge about the route or similar situations.

1.12   WHAT IS SYSTEMIC LEARNING?

As we have discussed above, in a dynamic scenario the role of environment and the interactions of the learning agent with environment become more and more important. Interestingly, it is an important thing to determine environment boundaries and understand the rewards and penalties of any action with reference to environment. As it becomes more and more complex and becomes more and more difficult, it also becomes very important to define the environment in dynamic scenarios. Furthermore, it even becomes necessary to understand the impact of any action from a holistic perspective. The sequence of percepts with reference to a system may need to be considered in this case. That makes it necessary to learn systemically. The fact is that sometimes rewards may not be immediate while it might be necessary to take into account the system interactions with reference to an action. The rewards, penalties, and even the resultant value need to be calculated systemically. To exhibit systemic decision making, there is a need to learn in a systemic way. The capture and building percept with all system inputs and within system boundaries is required.

Learning with a complete system in mind with reference to interactions among the systems and subsystems with proper understanding of systemic boundaries is systemic learning. Hence the dynamic behavior and possible interactions among the parts of a subsystem can define the real rewards for any action. This makes it necessary to learn in a systemic way.

1.13   WHAT IS SYSTEMIC MACHINE LEARNING?

Making machines to learn in a systemic way is systemic machine learning. Learning in isolation is incomplete—furthermore, there is no way to understand the impact of actions on environment and long-term prospects of reaching a goal. But the other aspect of systemic machine learning is to understand the system boundaries, determine the system interactions, and also try to visualize the impact of any action on the system and subsystems. The systemic knowledge building is more about building holistic knowledge. Hence it is not possible with an isolated agent, but rather it is the organization of intelligent agents sensing the environment in various ways to understand the impact of any action with reference to environment. That further leads to building the holistic understanding and then deciding the best possible action based on systemic rewards received and inferred. The system boundaries keep changing and the environment function in traditional learning fails to explore in multiobjective complex scenarios. Furthermore, there is a need to create the systemic view, and systemic machine learning tries to build this systemic view and make the system learn so that it can be capable of systemic decision making. We will discuss various aspects of systemic learning in Chapters 2 and 3.

1.14 CHALLENGES IN SYSTEMIC MACHINE LEARNING

Learning systemically can solve many real-life problems available at hand. But it is not easy to make machines learn systemically. It is easy to develop learning systems that work in isolation, but for systemic learning systems there is a need to capture many views and knowledge about the system. For many intelligent systems just based on percept or rather a sequence of percepts, it is not possible to build a system view. Furthermore, to solve the problems and simplify the problems to represent a system view, there is a need to go ahead with a few assumptions, and some of these assumptions do not allow us to build the system view in the best possible way. To deal with many complexities in systemic machine learning, we need to go for complex models; and in the absence of knowledge about the goal, the decisions about the assumptions become very tricky.

   In systemic thinking theory the cause and effects can be separated in time and space, and hence understanding impact of any action within the system is not an easy task. For example, in the case of some medicine we cannot see the results immediately. While understanding the impact of this action, we need to decide time and system boundaries. With any action the agent changes its state and so does the system and subsystem. Mapping these state transitions to the actions is one of the biggest challenges. Other challenges include limited information, understanding and determining system boundaries, capturing systemic information, and systemic knowledge building. In subsequent chapters the paradigm of systemic learning with the challenges and means to overcome them are discussed in greater detail.

1.15 REINFORCEMENT MACHINE LEARNING AND SYSTEMIC MACHINE LEARNING

There are some similarities between reinforcement learning and systemic machine learning while there are subtle differences. Interestingly, reinforcement learning and systemic machine learning is based on a similar foundation of exploration in a dynamic scenario. Furthermore, reinforcement learning is still more goal centric while systemic learning is holistic. The concept of systemic machine learning deals with exploration, but more thrust is on understanding a system and the impact of any action on the system. The reward and value calculation in systemic machine learning is much more complex. Systemic learning represents the reward from the system as the system reward function. The reward it gets from various subsystems and the cumulative effect is represented as the reward for an action. Another important thing is inferred reward. Systemic machine learning is not only exploration, and hence the rewards are inferred. This inference is not limited to the current state, but it also inferred for n states from the current state. Here n is the period of inference. As the cause and effects can be separated in time and space, rewards are accumulated across the system and inferred rewards are accumulated from the future states.

1.16 CASE STUDY PROBLEM DETECTION IN A VEHICLE

As discussed in great detail in the next chapter, a system consists of interrelated parts that together work to create value. A car is a system. When there is startup trouble in a car, it is advised to change the ignition system. In reinforcement learning, you change the ignition and the car starts working fine; after 8–10 days the car again starts giving the same problem. It is taken to mechanic who changes an ignition system again. This time he uses an ignition system of better quality. The issue is resolved and you receive positive reward. Again after a week or so the car begins giving startup trouble once again. Taking into account the whole system can help to solve these types of problems. The central locking system that was installed before this problem occurred is actually causing the issue. The impact on the whole system due to the central locking system is not considered previously, and hence the problem remains unattended and unresolved. As we can see here, the cause and effects are separated in time and space and hence no one has looked at the central locking system. In systemic machine learning, considering the car as a system, the impact of central locking is checked with reference to a complete system, that is, complete car and hence the problem can be resolved in a better way.

1.17 SUMMARY

Decision making is a complex function. Day by day the expectations from the intelligent systems are increasing. Isolated and data-based intelligence can no longer meet expectations of the users. There is a need to solve the complex decision problems. To do this, there is a need to exploit the existing knowledge and also explore new routes and avenues. This happens in association with environment. For any action the environment provides the reward. The cumulative reward is used in reinforcement learning to decide actions. Reinforcement learning is like learning with critic. Once an action is performed, a critic criticizes it and provides feedback. Reinforcement learning is extremely useful in dynamic and changing scenarios such as boxing training, football training, and business intelligence.

Although reinforcement learning is very useful and captures the essence of many complex problems, the real-world problems are more systemic in nature. Furthermore, one of the basic principles of systemic behavior is that cause and effects are separated in time and space. It is very true for many real-life problems. There is a need of systemic decision making to solve these complex problems. To make systemic decisions, there is a need to learn systemically. Systemic machine learning involves making a machine learn systemically. To learn systemically, there is need to understand system boundaries, interaction among subsystems and impact of any action with reference to a system. The system impact function is used to determine this impact. With broader and holistic system knowledge, it can deal with complex decision problems in a more organized way to provide best decisions.

REFERENCE

1. Senge P. *The Fifth Discipline—The Art & Practice of The Learning Organization.* Currency Doubleday, New York, 1990.

CHAPTER 2

Fundamentals of Whole-System, Systemic, and Multiperspective Machine Learning

2.1 INTRODUCTION

As we have discussed in Chapter 1, learning refers to mathematical representation and extrapolation of data and experiences based on input and output mapping. It is generally data centric, and these data are either pattern or event based. Here an event is a single occurrence of incidence used for learning while pattern refers to repetitive occurrences of similar events. The event has attributes and these attributes are used for learning.

Learning is generally bounded by local boundaries of reference. Typically these boundaries define region of effectiveness of a system. The samples from this region used for learning are referred to as learn sets. The learn sets that are used for training the system are generally a representation of the perceived decision space. The decision making is confined by local boundaries. An important question to be asked is, What should the search space be, and ideally where should these boundaries be? Understanding the relevance of information with reference to a decision problem is a complex and tricky task.

The concept of systemic decision making is based on considering the system while making decisions. Systemic decision making refers to system boundaries rather than local boundaries restricted by local information. Systemic learning means learning from a systemic perspective. Systemic learning deals with learning with reference to a system and takes into account different relationships and interactions with the system to produce the best possible decisions. It takes into account historical data, pattern, and old events similar to the present situation and behavior of relevant systems and subsystems. Apart from that, it considers the impact of any decision on other system components and interactions with the other parts of the system.

This chapter discusses the need for systemic learning and selectively using the learned information to produce the required results. The systemic learning tries to

Reinforcement and Systemic Machine Learning for Decision Making, First Edition. Parag Kulkarni.
© 2012 by the Institute of Electrical and Electronics Engineers, Inc.
Published 2012 by John Wiley & Sons, Inc.

capture the holistic view for decision making. Traditional learning systems are confined to data from events and space limited by an actual visible relationship. If we analyze systemic learning in detail, it leaves some fundamental aspects required in learning untouched and tries to build a new paradigm. Whole system learning with selective switching tries to get the best of both worlds. "Whole system learning" combines the concepts of traditional machine learning along with system thinking, system learning, systemic learning, and ecological learning. The most important part of this learning is about knowing or rather understanding the system, subsystems, overlapping among different systems, and interactions among them. This is determined based on the areas of impact, interactions, and points of impacts. The most important part of this learning is determining and emphasizing on the highest leverage points while making any decision or guiding any action. Here the highest leverage points refer to the time and the decision point that can lead to the best outcome. The positive and negative expected behavior with reference to these decision points is one important aspect of it. For example, in case of acupressure, one needs to apply optimal pressure at particular highest leverage points. Even in the case of other medicines, the effect of the medicine also depends on when it is given. Furthermore, these highest leverage points keep changing with reference to scenarios and context. The learning should enable the locating of highest leverage points in changing and dynamic scenarios.

Systemic learning includes the analysis of different dependencies and interdependencies and determining these leverage points dynamically. Another important aspect of learning is working on these leverage points. This chapter introduces the concept of selective systemic and whole systemic learning and further implementation of it in the real-life scenarios.

To make systemic learning possible, we need to have complete information about the system. For this purpose we should have decision-centric analysis of the system. To make this possible, we need to learn from multiple views and perspectives. The learning from particular or only visible or available perspectives could build incomplete information. In the absence of information and knowledge from all perspectives, systemic decision-making could be a very difficult task.

2.1.1   What Is Systemic Learning?

Systemic learning is learning that takes into account a complete system, its subsystems, and interactions before making any decision. This information can be referred to as systemic information. Systemic learning includes identification of a system and building systemic information. This information is built from the analysis of perspectives with reference to systemic impact. This learning includes multiple perspectives and collection of data from all parts of system. Furthermore, it includes the data and decision analysis related to impact. Decisions are part of a learning process, and the learning takes place with every decision and its outcome. The knowledge augmentation takes place with every decision and decision-based learning. This learning is interactive and driven by environment, which include different parts of the system. The system dependency of learning is controlled and is specific to the problem and system.

Systemic learning is inspired from systemic thinking. Systemic learning is about understanding systems, subsystems, and systemic impact of various actions, decisions within the system, and decisions in a systemic environment. It is more about learning about the actions and interactions from a systemic perspective.

Figure 2.1 tries to highlight the difference between systemic and analytical thinking.

Figure 2.1 tries to depict the relationship between analytical thinking, systemic thinking, and whole-system learning. System thinking includes analytical and synthetical thinking. The logical mapping along with inferencing, when combined with systemic thinking and other aspects, builds a platform for whole system learning. Synthetical thinking deals with observation and facts and combining separate element as a whole body. Synthetical thinking is thinking in terms of whole system based on facts and observations. It focuses on interactions among the parts of the system. The search space, decision space, and action space is what we generally focus on for decision making. System space is space confined by decision impact and dependencies.

If we want different outcomes from a situation, we have to change the system that underpins the situation in such a way that it delivers different outputs. Systemic thinking finds and focuses on the theme across the elements, while analytical thinking selects and focuses on the most attractive or promising element. Interestingly, to have a better and sustainable outcome, both types of thinking are necessary. But in some scenarios they lead to conflicting decisions.

Figure 2.2 depicts the characteristics of analytical thinking and systemic thinking. Analytical thinking allows one to select an element, while systemic thinking is about

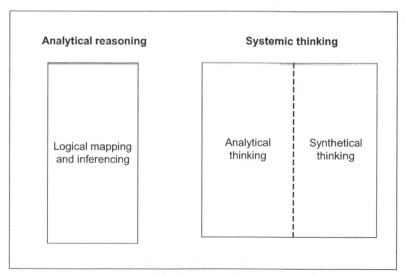

Figure 2.1 Systemic learning view.

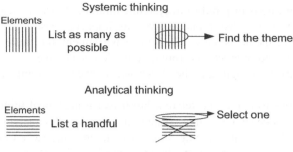

Figure 2.2   Systemic thinking.

finding a theme. Analytical thinking can be extended to a pattern, while a theme has many more dimensions and it does not point to a decision but builds a guideline for decision making. Systemic learning includes systemic modeling, systemic problem solving, and systemic decision making.

In real life we have always limited information, limited view, and fragmented picture available of the event or series of events. The information available to us is biased by our own thinking or biased by the perspective—in short the information made available to us is collected, interpreted, mined, and/or derived from a particular perspective. As a result, this information gives us a limited view. Hence decisions taken from this information cannot take into account the impact of these decisions beyond the limits of view available. Since there is an incomplete view, the impact of these decisions are always studied from a particular perspective and within those visible limits. Without availability of comprehensive information the systemic decision-making is not possible. In short, systemic decision making is not possible in absence of systemic learning. Hence systemic decision-making demands systemic or comprehensive information building.

Systemic learning is about building abilities in learning to make systemic decision making possible. The real-life problems are complex and all actions are interdependent. Furthermore, the visibility of the impact of action or a decision is confined by the system view and time view boundaries. The purpose of the system learning is to extend the time view and system view boundaries with learning and inference. Systemic learning works like an expert where the expert is in better position than others when it comes to decision making in his own area than the person without proper inferencing skills and required knowledge beyond the visible facts.

2.1.2 History

System thinking is not new, and researchers have been working on this concept from centuries. In Indian, Chinese, and Egyptian philosophies we find the mention of system thinking and the use of this paradigm by philosophers. System thinking is developed or rather presented from the management perspective, and the book *Fifth Discipline* by Peter Senge made it popular as a proven management technique [1].

This is further used to develop many management and decision-making tools. Systemic learning is about using systemic thinking, while learning and systemic machine learning is about systemic learning applied to machine learning. Systemic learning is not a new field, and this chapter provides various developments in systemic learning and also traditional machine learning methods that can be used to make systemic machine learning possible.

2.2 WHAT IS SYSTEMIC MACHINE LEARNING?

The best way to define systemic machine learning is as follows:

> Systemic machine learning is the learning that empowers machines' intelligence to make systemic decisions (based on learning from experiences)

Here machine learning is about giving abilities to machines to learn from experiences to enable them to make decisions in complex, not so complex, and new scenarios. Machine learning movement is to make the machine learn from experience. This experience comes in the form of historical data, historical examples represented in data series, and feature vectors. In machine learning, historical information and trends are used for learning and decision-making.

In case of systemic machine learning, it is necessary to understand and define the system space for the decision problems. The data used for learning needs to represent whole-system features. The systemic machine learning uses the historical information, data, and inferencing to empower machines to make systemic decisions. The data about the whole system is generally not available, and also the analysis of the system and interaction among its different parts are not visible. In absence of this information the machine needs to develop the capability to acquire this information and build knowledge based on available information. Machines also need to incrementally upgrade their knowledgebase to acquire a better position for decision making. This is holistic machine learning with the use of information and inferred impact beyond view boundaries.

The decision maker's view is depicted in Figure 2.3. A decision maker can view only some part of the overall system. There are some subsystems or parts of subsystems that the decision maker could not view. Similarly, some impacts of decision such as those depicted with dotted lines are not visible to the decision maker. The view is restricted by space and time, and to infer or to visualize the nonvisible impact in time and space beyond the view is the challenge in front of system learning.

Figure 2.4 depicts a system impact diagram assuming the car as a system. All parts of the car have dependencies or relationship with other parts. Sometimes it is visible, but in other cases it is indirect. As you can see from this example, dependencies are among the different parts.

To learn about the car and repair or resolve the problem, the mechanic must have the systemic view because the change in one part may have some impact on other parts. For example, when I installed central locking in my car, my car has began giving

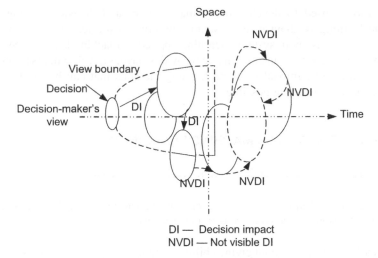

DI — Decision impact
NVDI — Not visible DI

Figure 2.3   Systemic view diagram.

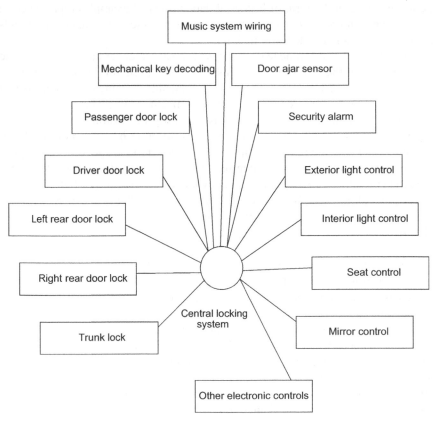

Figure 2.4   Systemic decision-making states.

starting trouble and mechanics took 2 days to identify the issue. Typical dependencies in case of a central locking system are depicted in Figure 2.4. These types of issues generally result due to the absence of a system view, lack of knowledge of mechanics, and the inability to associate cause and effects that are separated in time and space. Thus the concept of system space is very important. System space is an active region one needs to consider for optimal systemic learning. In the above example it can be a complete car or some sets of circuits in the car.

2.2.1 Event-Based Learning

Event-based learning is one of the basic forms of learning. It is a special case of learning where learning space is an event. All learning models and inference are based on single event in this case. Occurrence of a particular event in supervised learning may sometimes act as a decision parameter. Also, that event may play an important role in building decision scenarios. It is the use of some significant occurrence to achieve the learning objective and later in decision making. To avoid wrong decisions that may result because of an event-based learning, we need to ensure that it should be based on patterns. Although this may be the case, event-based learning cannot be discarded always and also pattern-based learning cannot be always the best solution. So the systemic learning is not forcing a decision mechanism, and it allows selecting the decision making on the merit of a problem.

In event-based learning as depicted in Figure 2.5 the system learns from the event and its outcome that learning drives the decision making. This learning may lead to issues when the event is not representing a real problem or behavior of the decision space.

In *pattern-based learning* instead of an event, the patterns that are derived from the outcome of a series of events drive decision making. The pattern-based learning overcomes some of the shortcomings of event-based learning and decision making since it is not driven by a single event. The repeated occurrence of a similar event or statistical pattern is used for learning. In this case the pattern is assumed to represent the behavior of decision space. Typical pattern-based learning is depicted in Figure 2.6.

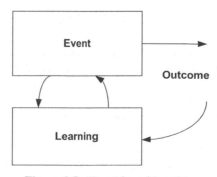

Figure 2.5 Event-based learning.

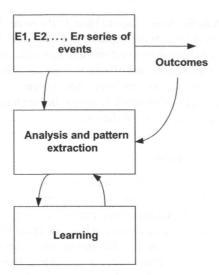

Figure 2.6   Pattern-based learning.

Structural Learning: Structure-based learning uses information patterns, but the learning is based on a framework that results from system structure. In systemic learning, pattern-based, event-based, and structure-based learning are used together. All pattern-based and event-based information matches to structural input for systemic learning.

Systemic machine learning makes use of historical information to identify patterns; it identifies a system and interactions among its different subsystems and system structures. It reinforces decisions, with mapping based on occurrences of new events. The systemic learning uses knowledge about a system and systemic impact of inferred learning based on prior learning. This should include inferencing on time axes. Systemic learning works in system space, and the decision is theoretic.

2.3   GENERALIZED SYSTEMIC MACHINE-LEARNING FRAMEWORK

Figure 2.7 provides a framework for systemic machine learning. System structure refers to various inbuilt assumptions, and learning and understanding those assumptions is the important part of system learning. In this framework, systemic machine learning uses historical data, system identification or system structure, system interaction study, and perspective understanding as input. This information is used to identify the highest leverage points in system space. Inference engine uses the above analysis to provide impact analysis. The system uses decision scenarios and decision perspective and impact analysis to build a decision matrix, which is the outcome of learning. The decision matrix is used to provide the solution, but in

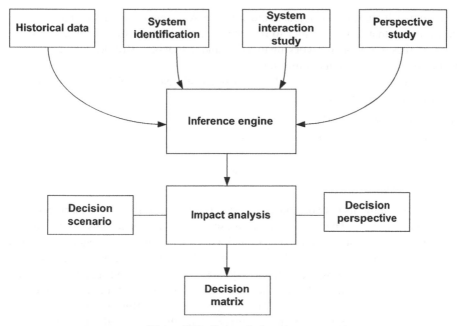

Figure 2.7 Systemic learning.

systemic decision making the process of learning is continuous and the process of decision making is interactive. In continuous learning the system structure and impacts are probed again and again to form the decision matrix. This interactive learning allows the learning system to identify the best decision and with reference to the highest leverage points.

Explanation: This framework tries to capture system interactions and determine the decision matrix with reference to the highest leverage points. Interestingly, the highest leverage point has the location and the time aspect. It is about the time when the decision is to be made and about the system point where it should be applied. The decision matrix represents and takes into account all these factors along with decision perspectives and scenarios that help this framework in determining the optimal decision. The impact analysis block is responsible for interactions with the environment and system and produces the decision matrix.

The question remains whether the system space we are referring to should be fixed or very specific to a decision problem. In the next section, we will discuss the system and system-space-related aspects.

2.3.1 System Definition

In real life many times it seems very obvious to define a system. But actually it is much more complex than what it looks. The system has a purpose for which all its parts work together. In short, a complete system works together for the purpose. The broad system may include every object in the universe. A system is a combination of various components that work together to perform a particular function that is not possible without these components working together and cannot be performed effectively even if any of the components is missing.

To reduce the complexity and from a practical analysis of decision impact, the system can be defined as follows:

Any organized assembly of resources and procedures united and regulated by interaction or interdependence to accomplish a set of specific functions.
Or
A collection of personnel, equipment, and methods organized to accomplish a set of specific functions.

Any two elements with dependence ratio greater than d from the scenario S belong to subsystem S_s. The union of all these subsets forms a subsystem. All the subsystems are a subset of the system based on prominent dependencies and a general decision scenario. In the case of very low dependencies of the particular subsystem on other subsystem, the subsystem forms the system itself. The system boundaries are determined by dependencies and interactions. In the real-life scenario the whole world is a system itself. This is true in all cases. But to simplify and make mathematical representation of the system possible, there is a need to exclude parts or regions with lower dependencies. This not only makes systemic learning possible but also makes it effective.

A scenario or object S belongs to subsystem/system S_s if the dependence ratio $> d$

Here dependence ratio represents the relationship of the object with system. This dependence ratio helps us in deciding the system space. Another aspect of a percept sequence is that it is time dependent. Many times the impact or outcome is visible in percept that is separated in time.

There is the possibility that cause and effect are separated in time. The outcome at different time instance allows inferencing when we are dealing with cause and effects that are separated in time. Figure 2.8 depicts the subsystem impact and outcomes over the time T_{-0} to T_{-n}. The percept sequence from T_0 to T_n helps in giving the view of the system space that can help in inferencing the system behavior.

These patterns and interdependence evolve or can be realized over the period of time. That is the reason why patterns keep changing. Furthermore, the impact relationship and the visibility of cause and effect are separated over the time and need to be tracked. The systemic learning system needs to capture this.

The cause-and-effects relationship needs to be mapped. A decision in isolation without understanding its impact out of the view boundaries may not yield desired results. A typical example of cause and effect over the time is depicted in Figure 2.9. In this example, because there is a shortage of apples, the apple production is promoted. That results in excess apples in the market. That makes farmers sell apples at very low rate, resulting in their monetary loss. This makes them reluctant to go for apple production for subsequent years, and ultimately the decision to increase production of apples results in a shortage of apples.

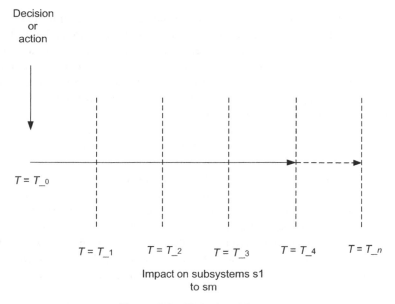

Figure 2.8 Time dependence.

2.4 MULTIPERSPECTIVE DECISION MAKING AND MULTIPERSPECTIVE LEARNING

Multiperspective learning is required for multiperspective decision making. Here multiperspective learning refers to learning from the knowledge and information acquired and built from a different perspective. The multiperspective learning process

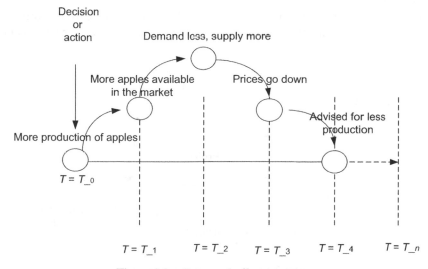

Figure 2.9 Cause and effect over time.

includes methods to capture perspectives and the captured data, information, and knowledge from different perspectives. The multiperspective learning builds knowledge from various perspectives and represents it so that it can be used for a decision-making process. The perspective refers to the context, scenario, and situation that impact the way we are looking at a particular decision problem. An intelligent agent captures a sequence of percepts. These sequences are separated in time scale. Multiple agents can capture percepts separated in feature space.

In Figure 2.10, P_1, P_2, ..., P_n represent different perspectives. Each of these perspectives is represented as a function of features. These perspectives may have overlap among them. There can also be some feature overlap. In some cases the feature may be the same but the overall weight and representative values may be different. The feature difference can be there because some features that are visible from a particular perspective may not be visible from other perspectives. The representative feature set should contain all the possible features.

As per the definition, perspective is the state of one's ideas, the facts known to an individual, and so on, that have a meaningful interrelationship. It is about seeing all the relevant data for a particular problem space in a meaningful relationship from the available window.

The perspective-based information can be represented as an influence diagram. This influence-diagram–based representation can help us to get the context of the decision-making right. Setting the context and structuring decision objectives and iterating again and again to finalize the context for reinforcing the same will lead to better systemic learning and better understanding of system space. An influence diagram is a graphical representation of the decision (scenario) situation. There can be other ways to represent decision situations and relationships. We have chosen influence diagrams because they can help us in most appropriately representing system relationships and also it is very simple and less complicated way of representation.

In learning, the information presented is generally made available from a particular perspective. But in a real-life scenario for a simple problem, there can be a number of

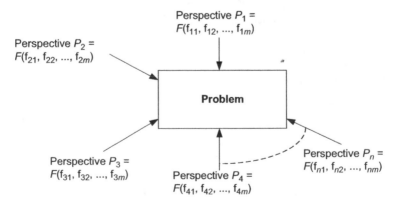

Figure 2.10 Multiperspective learning.

perspectives possible. Some of the perspectives can be directly derived from the objectives. These play the major role in analytical thinking and analytical decision making. The multiperspective learning includes information capturing from different perspectives. The decision-making perspective is many times different than the learning perspective. Although in some cases it is matching, the data from the other perspective is missing and due to that the decision making may have some issues. The fundamental idea of multiperspective learning is to capture information from all possible perspectives. The information from the various perspectives is used to build the knowledge, and that knowledge is used for effective decision making.

The information that is most relevant from one perspective may not be that relevant or may not be relevant at all from another perspective. In such a scenario there is a huge gap between what is learned and what should be learned, and what looks like an obviously right decision may not yield suitable results.

Figure 2.11 depicts an influence diagram (ID) representation of a market scenario and relationship between marketing budget, product price, cost, and profit. The shapes and in some cases even colors are also used in an influence diagram for representation of an object. Influence diagrams show the relationships among objects and actions. These relationships can be mapped to probabilities, as we will see in the next section.

The same relationships can be very well represented by using a decision tree. A decision tree provides the transition or decision path based on measurement of some parameters, and as we go to the next level we are dealing with hierarchical decision making. The decision rules can be very well represented as decision trees. Figures 2.12 depicts a decision tree and Figure 2.13 depicts an influence diagram.

The influence diagram and decision tree are used to represent a different kind of information. The influence diagram shows dependencies among variables very clearly. In a semiconstrained influence diagram the possibility of dependency is shown. Some examples of an influence diagram with perfect information, imperfect information, and no information are shown in Figures 2.14–2.16.

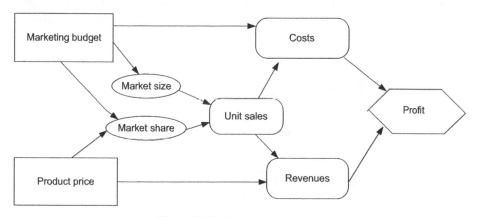

Figure 2.11   Influence diagram.

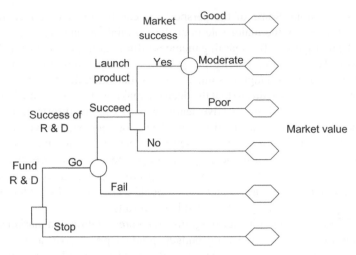

Figure 2.12   Decision tree.

Actually, in system learning, having all the information is not possible and most of the time case 2—that is, imperfect information—is obvious. There is need to model and acquire systemic information in the case of imperfect information.

Figure 2.14 represents a scenario with no information, while Figure 2.15 depicts a scenario with complete information. In real life there is no situation, where complete information is available, hence it is a hypothetical scenario.

The typical influence diagrams are shown in Figure 2.16. Influence diagrams are particularly helpful when problems have a high degree of conditional independence, when a compact representation of extremely large models is needed, when communication of the probabilistic relationships is important, or when the analysis requires extensive Bayesian updating. The conditional independence allows us to represent conditional probability in a more useful way for decision-making problems and hence is very important for machine learning. Influence diagrams represent the relationships between variables. These relationships are important because they reflect the analyst's or the decision-maker's view of the system.

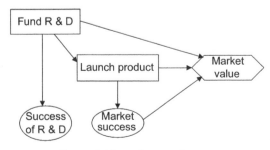

Figure 2.13   Influence diagram.

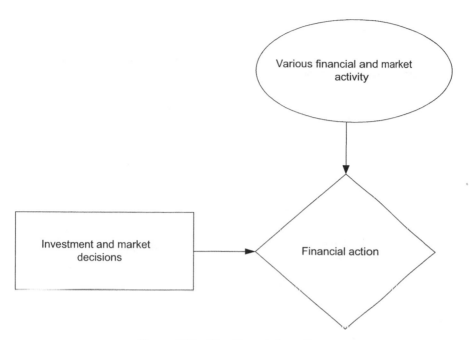

Figure 2.14 ID with no information.

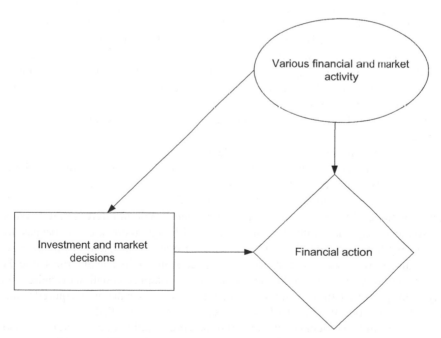

Figure 2.15 ID with perfect information (hypothetical scenario).

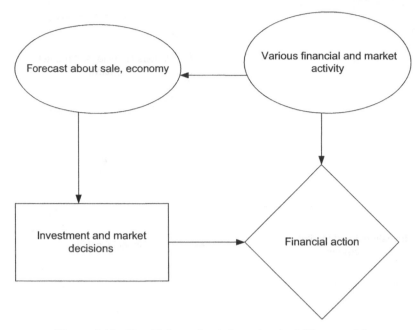

Figure 2.16   ID with imperfect information (real-life scenario).

In short, a probability influence diagram is a network with directed graphs but no directed cycles. As we move to subsequent chapters, we will look into various representations and its use in modeling and representing various decision scenarios in real-life problems—that is, with imperfect and partial information.

The perspective-based information can be represented as an influence diagram. As already explained in this chapter, a "Bayesian decision-making" influence diagram is a representation of the influence of different inputs resulting in a state transition which is also represented with transition probabilities.

Here influence diagrams are associated with probabilities of occurrence of an event (Figure 2.17).

This feature may also facilitate knowledge acquisition. It represents the overall decision scenario. For example, a clinician expert may be able to assess the prevalence of disease and sensitivity and specificity of a diagnostic test more easily than she could assess the post-test probability of disease. After the influence diagram is drawn to facilitate probability assessments, all updating and Bayesian inference are handled automatically by the evaluation algorithms. Although there are approaches for performing Bayesian updating within a decision tree, for problems with extensive Bayesian updating, such as sequential-testing decisions, influence diagrams ease the burden on the analyst by reducing the need for complex equations required for Bayesian updating in the tree. Influence diagrams also reduce the time required to find errors that may be introduced when these equations are specified.

Here we will use influence diagram (ID) for representation of a decision scenario. In a real scenario an influence diagram is the representation of the part of the system

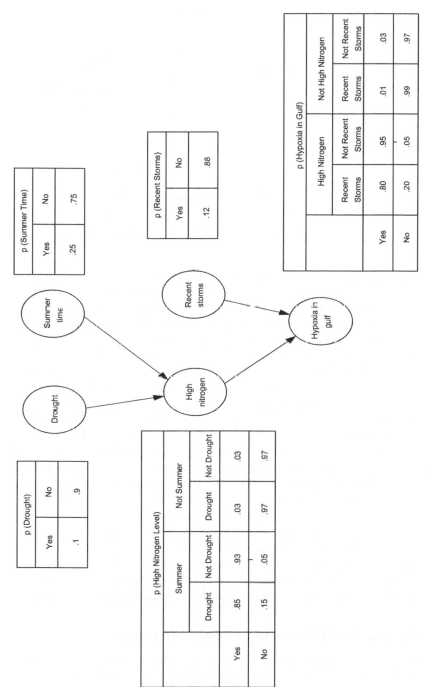

Figure 2.17 Examples with probability.

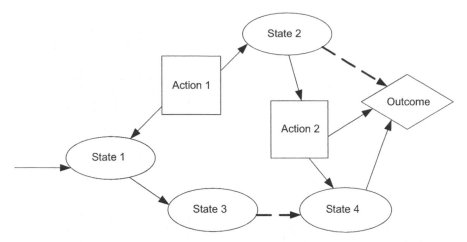

Figure 2.18   Partial decision scenario representation diagram—PDSRD.

that is visible to the decision maker. We can refer this as perceived decision boundaries. Also it can be a system representation from a particular perspective. In real life it is always possible that even the complete information from the obvious perspective or decision-maker's perspective is not available at the time of making decision. This limited information about dependency as well as insufficient information for decision making can lead us to represent decision scenarios in a slightly different way and that we will refer to as a semiconstrained influence diagram (SCID), also called a partial decision scenario representation diagram (PDSRD).

PDSRD has the relationships represented in a fuzzy manner. These PDSRD fuzzy relationships become concrete as we combine more and more perspectives and the revelation of systemic information over the period of time (Figure 2.18).

The limited information available about decision scenarios, dependencies, and especially from a particular perspective or rather fragmented information or picture will be represented as a partial decision scenario representation diagram. Even PDSRD can be viewed as an influence diagram with constraints and limited information. A typical PDSRD is depicted in Figure 2.19.

Dashed lines appearing in the diagram indicate the possible relationship. In PDSRD there can be some relationships you are not very sure about. Some of the relationships are fuzzy and represented with the question marks on the link. The probabilities of transition are known for a few relationships while are not known for others. This helps in forming a partially filled decision matrix with fuzzy values.

2.4.1   Representation Based on Complete Information

Here complete information means all the systemic parameters available. This helps us determine all probabilities of transitions. Due to this the decision making becomes much easier. But in real life the complete information is not available at any point of time. When we have a complete picture available or we are absolutely sure that

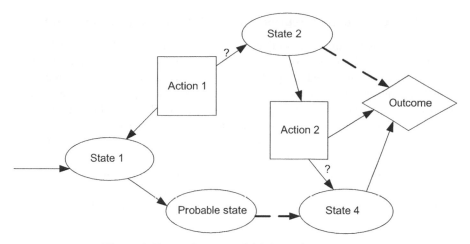

Figure 2.19   PDSRD—partial information diagram.

event- or pattern-based decision making can very well be used to solve the problem, this representation is used. Actually an influence diagram with complete information is a special case of that of PDSRD.

2.4.2   Representation Based on Partial Information

Generally there is only partial information available. This partial information can be represented using PDSRD. There are many such diagrams from different perspectives, and those in isolation cannot guide us to decisions. The representative diagram for all these representations is required for it. Representative Decision Scenario Diagram (RDSD) is the representation of decision scenarios by combining different PDSRDs. The representative Decision Scenario Diagram is a representation of multiperspective learning. It is actually the representation of knowledge acquired from all perspectives.

2.4.3   Uni-Perspective Decision Scenario Diagram

The PDSRDs are generally used to represent decision scenarios that are uni-perspective influence diagrams. The transitions and the probabilities associated with those transitions in this diagram represent decision-maker's perspective. Even ID with probabilities can be viewed as uni-perspective Decision Scenario Diagrams.

2.4.4   Dual-Perspective Decision Scenario Diagrams

To overcome the limitations of uni-perspective Decision Scenario Diagram, we represent information in dual-perspective decision scenario diagram. Here in a single diagram there are two probabilities, and transition patterns based on perspectives are represented. The dual-perspective influence diagrams can help in representing some of the not-so-complex problems where two perspectives can cover the most of the system and decision space.

2.4.5   Multiperspective Representative Decision Scenario Diagrams

As there are many possible perspectives in real-life complex problems and a decision needs to be made after taking into account these perspectives, there is need for multiperspective IDs. As discussed in a previous section, partial decision scenario representation diagrams represent different perspectives where a single partial decision scenario representation diagram represents a particular perspective. There are PDSRDs for each perspective. These PDSRDs are used to form a representative Decision Scenario Diagram (DSD) for a particular decision scenario. The representative DSD is used for decision making and allows multiperspective decision making.

In the case of no knowledge of dependency from a particular perspective, the representative DSD will not represent that particular perspective. More and more information with reference to perspectives is incorporated in representative DSD.

2.4.6   Qualitative Belief Network and ID

Bayesian belief network or influence diagrams acknowledge the usefulness of these frameworks for addressing complex, dynamic real-life problems. The large number of dependencies needs to be represented in the form of probabilistic relationships. At any point of time the representation of an expected event logically for decision making is in the form of conditional probability. The usually large number of probabilities, representation of relationships, and utilities are required for their application and problem solving. However, it is often difficult to map and represent these dependencies. The belief-network framework in itself does not provide decision making under uncertainty, as decision making involves not only knowledge of the uncertainties in a problem under study but also knowledge of the decisions that are at a decision-maker's disposal and of the desirability of their uncertain consequences. This even has perspectives and context in hidden form. This makes it necessary to have proper and close-to-perfect information for decision making. The framework of influence diagrams is tailored to decision making [2]. The influence diagram combined with belief network can be opted to enhance belief network. It can provide the knowledge capturing and knowledge augmentation mechanism.

Qualitative belief networks, introduced by M. P. Wellman as qualitative abstractions of belief networks, bear a strong resemblance to their quantitative belief networks in many ways [3]. A qualitative belief network comprises a graphical representation of the independences among a set of statistical variables, once more taking the form of an acyclic digraph. Instead of conditional probabilities, however, a qualitative belief network associates with its digraph qualitative probabilistic relationships. These interdependencies can further be extended with systemic relationships.

Qualitative influence diagrams are qualitative abstractions of influence diagrams. A qualitative influence diagram, like quantitative counterpart, comprises a representation of the variables involved in a decision problem along with their interrelationships, once more taking the form of an acyclic digraph. Instead of conditional probabilities, however, a qualitative influence diagram encodes qualitative influences and synergies on its chance variables. Instead of utilities, it specifies qualitative preferential relationships. These preferential relationships capture the preferences of the decision maker and, hence, pertain to the diagram's value node. These preferred relationships can be used to represent the partial information when it comes to PDSRD.

2.5   DYNAMIC AND INTERACTIVE DECISION MAKING

Every decision and every action lead to more revelations. With time, more and more information about the system becomes available. This new information even builds a new perspective for decision making. To give the best decisions or rather systemic decision, there is need of dynamic and interactive learning and decision making. The new information changes the decision scenarios, and the system needs to have differential learning capability. Systemic learning needs to be dynamic and interactive. Here dynamic means it will be able to accommodate new decision scenarios that arise due to systemic interactions and interactive learning, which interacts with the system and thus builds and updates knowledge. The new information and knowledge produced/built is used for learning. The dynamic decision making demands the ability to adapt to changing decision scenarios and even features.

2.5.1   Interactive Decision Diagrams

Interactive decision diagrams can be used to represent interactive learning scenarios. Interactive decision diagrams allow recursive nesting of decision-making scenarios.

Interactive dynamic decision diagram is a generalization of dynamic decision diagrams. They help in computing finite look-ahead approximation. These interactive decision diagrams can be applied for systemic learning and interactive decision making. Here the interaction of the decision scenario with environment and system space is represented.

2.5.2   Role of Time in Decision Diagrams and Influence Diagrams

There are at least three particular cases when the value of time or ordering in time play an important role. Time is not explicitly declared. If the decision problem, modeled by an influence diagram, has no periods of time clearly stated, a diagram is constructed sequentially: At first chance, variables and dependency arcs between them are introduced, then decision variables with information links are added, and next utility nodes are defined and connected with other nodes. The ID is ready, but before using it we have to term its realization: an attachment of functions to the appropriate variables. This means that the chance nodes and variables are associated with conditional probability functions (or prior probabilities for nodes without parents) and the utility nodes and variables are associated with utility functions. Decision nodes correspond to actions taken by external agents; ID defines information needed for each decision (by information arcs) and, sometimes, the order of decisions. An order depends on the structure of the ID and its interpretation.

The ID is assumed to be of static type and nature. ID can be time sliced in different ways. Time-sliced ID can be used to find best moments for decision making. Also the highest leverage points and moments are identified for decision making. Here we want the time-sliced ID to find out the moments for information augmentation and portray the dynamic scenarios those we do not want to miss especially in case of systemic learning. These "time-sliced IDs" allows selective enhancement, and also the representative influence diagram can be formed at the decision moment.

2.5.3 Systemic View Building

The system rules defined in *Systems Thinking* by Peter Senge suggest that a wrong decision today ("that probably we do not know is wrong because of the limited information and view available at this moment") can result in a bigger problem tomorrow [1]. In some cases the impact of the decision may not be seen in the immediate future (i.e., cause and effect are separated in time). What can help you to take better decision today is the capability to redefine system scenarios, parameters, and interactions in the light of new information. It is continuous learning and understanding the system.

Figure 2.20 depicts the flow chart for decision analysis. While the system view is built, decisions are also analyzed. The decision is analyzed based on the systemic impact of the decision. The decision impact needs to be analyzed in the future, as the

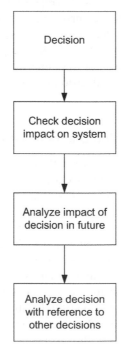

Figure 2.20   Decision analysis.

impact may not be visible today. Another important thing is the analysis of the decision with reference to other decisions in the system. These three analyses are referred to as

- System space boundary analysis
- System time boundary analysis
- Compatibility analysis

These analyses allow validating decisions and taking corrective measures. This analysis is also used while learning.

2.5.4   Integration of Information

The fragmented information available from various perspectives can be represented as PDSRD. There is need of integrating this information when we form a representative Decision Scenario Diagram (DSD). As the absolute integration of this information may not be relevant, we need to integrate it with reference to a particular decision scenario. The details of integration approaches are covered in subsequent chapters. During the integration process the SCIDs produced as a result of different perspectives are used. The inferencing along with other machine learning techniques are used during this process. The integration of information is carried out from bottom to up. The purpose of integration of information is to build a systemic view.

2.5.5   Building Representative DSD

The PDSRD are combined to form a representative DSD.

$$PDSRD_{\_1} = F(f_{11}, f_{21}, f_{31}, \ldots, f_{n1})$$
$$PDSRD_{\_2} = F(f_{12}, f_{22}, f_{32}, \ldots, f_{n2})$$
$$\vdots$$
$$PDSRD_{\_m} = F(f_{12}, f_{22}, f_{32}, \ldots, f_{nm})$$

The decision scenario decides weights of various features corresponding to different PDSRDs.

$$RDSD = (w_1 w_2 w_3 \ldots w_n) \times (\text{feature matrix})$$

The selective features are calculated.

The representative DSD is for a particular decision scenario. For a new decision scenario we will have a new representative DSD.

2.5.6   Limited Information

The limited or incomplete information is one of the major challenges in machine learning. The perspective-based formulation and overall perspective integration allows the inferencing of some missing data points to enable the decision making and learning based on limited or incomplete information. The available information in real-life problems can always have something missing, but the integration and inference allows us to build the required information scenarios. Furthermore, learning is a continuous process and as more data become available the inferred facts are further refined.

2.5.7   Role of Multiagent System in Systemic Learning

Use of various agents or rather multiagents can help in collecting systemic information. Homogeneous noncommunicating agents, heterogeneous noncommunicating agents, homogeneous communicating agents, and heterogeneous communicating agents are described in Table 2.1. The homogeneous noncommunicating agents are used to build the information for global versus local perspective in distributed environment. The communicating agents help in a better way to build systemic view.

Various agents build global perspectives, and the information about this is listed in Table 2.1.

These agents interact with the environment and with the sense status of the system. The adaptive agents dynamically probe the system. Figure 2.21 shows a representative agent that interacts with its environment to build domain knowledge and is responsible for actions.

Figure 2.22 depicts a multiagent system. These agents also interact among each other, which helps them in building better systemic view. The agents use the domain knowledge and feedback from the environment for learning.

A generic model of a learning step is depicted in Figure 2.23. The training function works on a number of hypotheses, and based on performance measures and solution quality it further reinforces the learning parameters. The interactions with systemic dependencies and environmental parameters help the system to learn systemically. The different systemic concepts explored in this chapter can be used for building learning model and framework.

Figure 2.24 depicts different systemic components in a document classification scenario. As the decision scope clarified and clarity about the perspective is sought,

TABLE 2.1   Agents and Cooperative Learning

Homogeneous Noncommunicating Agents	*Heterogeneous Noncommunicating Agents*
Local versus global perspective	Benevolence versus competitiveness
Different states	Social conventions
	Roles
	Modeling goals
Homogeneous Communicating Agents	*Heterogeneous Communicating Agents*
Distributed sensing	Understanding each other
Communicating contents	Negotiation
Mapping	Planning communication act
Group learning	Benevolence versus competitiveness
	Changing shape versus size
	Group learning

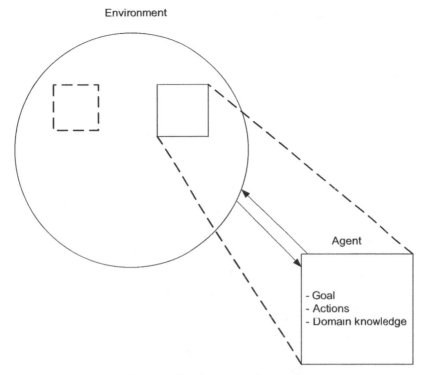

Figure 2.21   Agent-based system.

the systemic view can be generated. It uses various system components within defined system boundaries. But in real scenarios these decision boundaries will be defined by a decision scenario.

2.6   THE SYSTEMIC LEARNING FRAMEWORK

The systemic learning framework should be able to perform learning action including (a) determining system boundaries and (b) updating these boundaries based on different outcomes of interactions and impacts of decisions. It should be able to learn dynamically to adapt to continuously changing scenarios and deliver the best decision for the given decision scenario with reference to system space. The framework is expected to do the following functions:

System Detection: System detection refers to determining the system boundaries and different components and subsystems, which are part of the system. The system detection is also influenced by decision scenarios.

Mapping Systems: Mapping the system components based on impact and dependencies. This helps in decision making and validation of decisions.

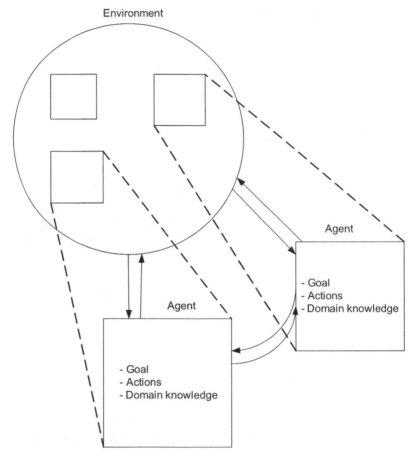

Figure 2.22   Capturing perspectives using multiagent system.

System Analysis: This refers to analysis of system and it should be continuous. The system needs to be analyzed in light of new information or outcome of the new decisions. The system analysis creates learning parameters for systemic learning.

Determine Interactions Among Subsystems: Another part of systemic analysis is determining interactions among different subsystems. These interactions help in building a decision matrix in a particular decision scenario.

Learning for Decision Impacts: The decision has impact on the system and the impact needs to be inferred in some cases. Learning of systemic decision impact is necessary to arrive at the right decision.

Perspective-Based System Impact Analysis: The information is generally incomplete and there is analysis based on a particular perspective. There is need of perspective-based impact analysis to choose the right balance among decision perspective in a particular decision scenario.

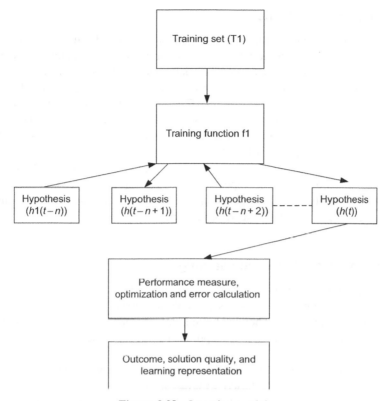

Figure 2.23   Learning model.

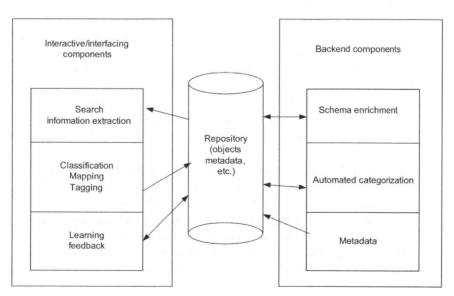

Figure 2.24   System components.

2.6.1 Mathematical Model

A generalized mathematical model for systemic learning is discussed in subsequent chapters. The mathematical model is based on defining the system and subsystems as a function of different sets of features. These subsystems interact with one another in a particular decision context. Perspective-based interactions among them is defined in terms of impact factors ($i1$, $i2$, ..., in). These impact factors from a particular perspective form the basis of PDSRD. PDSRD is defined as a matrix ($d1$, $d2$, ..., dn). Each PDSRD has a decision matrix that contains weight for a decision scenario. For a particular decision scenario all PDSRDs are combined to a form a representative DSD where every PDSRD decision matrix is defined based on weight for the decision scenario and decision matrix of respective DSD. The representative DSD is resolved in a particular analysis. These decision matrices and inference mechanisms for determining weights are at the heart of learning.

2.6.2 Methods for Systemic Learning

Systemic learning is the way of learning to determine the system impact and learning to empower the system to make the best possible decision in the interest of the system. For this it is necessary to infer beyond boundaries. Various methods can be used for it and can be optimized. Some of these approaches are listed below:

Object-Based Learning: Here the object refers to decision scenarios with data, and learning takes place based on these objects.

Fragmentation and Learning: Here the information is fragmented, and to get further clarity the information is again combined. The division and combination of information are done based on a case-to-case basis.

Multiperspective Learning: As discussed above, to learn from multiple perspectives, the learning is first done based on various perspectives. The learning based on various perspectives is combined to form a multiperspective decision matrix. The multiperspective decision matrix helps in making decisions.

Clustering at Various Levels: The learning at various levels can be supervised as well as semisupervised. The clustering at various view levels can be performed. These clusters formed at various view levels are used in the formation of decision matrix.

Subspace Clustering: The subspaces based on decision spaces and perspectives are formed. The subspace clustering based on limited prominent places is performed, and further integration clustering can be used.

Perspective-Based Incremental Clustering: Another important aspect is incremental learning; and as more and more information becomes available, even the perspective-based decision parameters vary. Perspective-based incremental clustering can be used to allow dynamic and incremental decision making.

2.6.3 Adaptive Systemic Learning

Adaptive systemic learning refers to selective systemic learning where the overall learning perspective is systemic, but it allows learning in different ways based on the problem. Figure 2.25 depicts the way in which adaptive systemic learning takes place. Another important aspect of adaptive learning is selective combination of more than one method and use of learned data simultaneously.

Machine learning theory also has close connections to issues in economics. Machine learning methods can be used in the design of auctions and other pricing mechanisms with guarantees on their performance. Adaptive machine learning algorithms can be viewed as a model for how individuals can or should adjust to changing environments. Moreover, the development of especially fast-adapting algorithms sheds light on how approximate-equilibrium states might quickly be reached in a system, even when each individual has a large number of different possible choices. In the other direction, economic issues arise in machine learning not only when the computer algorithm is adapting to its environment but also when it is affecting its environment and the behavior of other individuals in it as well. Connections between these two areas have become increasingly strong in recent years as both communities aim to develop tools for modeling and facilitating electronic commerce.

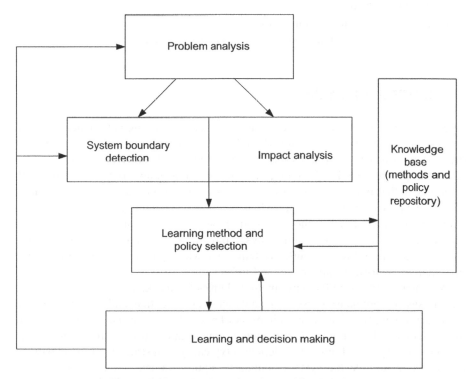

Figure 2.25 Adaptive systemic machine learning.

In adaptive systemic learning the following important components are considered:

- System interactions.
- Information and knowledge resides in the system.
- The learning algorithm is adaptive and chooses the method based on data availability and system.
- Selective use of methods and formation of feature vectors with prioritization weight.
- Changing learning parameters as per system status and learning requirements.
- Enhancing the abilities of learning with the revelation of more data and scenarios.
- Knowledge augmentation and selective use and knowledge mapping.

2.6.4   Systemic Learning Framework

The important components of a system-learning framework are as follows:

- System defining unit
- Understanding subsystems
- System interactions
- Analyzer for systemic impact of decision
- Decision selection based on multiperspective analysis

2.7   SYSTEM ANALYSIS

One of the most important parts of systemic machine learning is to understand and analyze the system. When learning is based on information restricted to a small part of the system or a particular subsystem the decisions and leanings inherently carry a particular perspective. Furthermore, the information available is also gathered with reference to that part of the system. Hence the information available is fragmented. Hence this poses many challenges when it comes to systemic decision making. To overcome these challenges, the information gathering and system analysis are two important parts. Based on system analysis, more and more information is gathered. Systemic learning tries to build systemic knowledge based on the fragmented information, historical knowledge, and inferencing.

A system interacts with environment and takes feedback. The learning is activity within system boundaries. System analysis tries to define and redefine system boundaries every time based on availability of new information. This analysis reveals the system structure. Figure 2.26 depicts a typical interaction of the learning with reference to a system. Figure 2.27 depicts a typical system structure. Typical parts of the structure include the system boundary, input parameters, output parameters, various subsystems, and environments. This structure also tells the relationships among these components of the system.

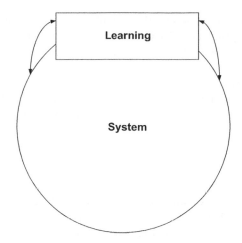

Figure 2.26   Learning component of the system.

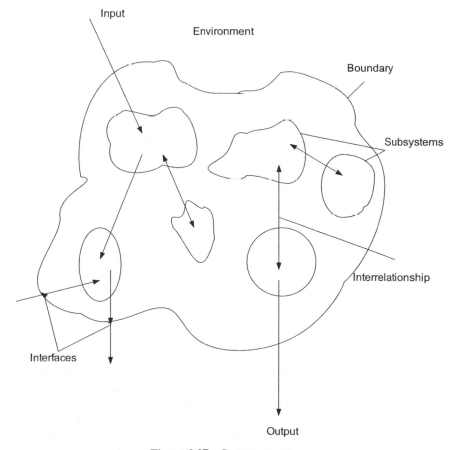

Figure 2.27   System structure.

Figure 2.27 depicts the system with a number of subsystems. There are various inputs coming to the system and also output given to the environment. Arrows depict the interactions among the system components.

Top-Down System Learning: In top-down system learning, the high-level system view decides the weights for the initial decision matrix. As the decision making progresses, we move down to the decision perspective.

Subsystem-Based System Learning: In this learning every subsystem is used for learning in isolation and integration is done at the time of decision making. System learning demands continuous improvement.

Learning Based on Desired Behavior: The supervised learning uses labeled data. Most of the times the learning is based on target value. The target value does not depict the system behavior. Target value may be maximizing the profit and the decision may lead to that, but system behavior may lead in a different direction. System learning is based on desired system behavior, and hence the objective function is formed based on desired system behavior.

Example of Systemic Machine Learning: In typical medical decision making where whole body is a system and hence before giving any medicine, the side effects of that medicine on other parts of the body are considered and hence the decision is systemic in nature. Hence all medical intelligent systems demand systemic machine learning.

2.8   CASE STUDY: NEED OF SYSTEMIC LEARNING IN THE HOSPITALITY INDUSTRY

In the hospitality industry various high profile hotels set their rates based on forecasting of demand. The forecasting of demand is calculated based on various patterns such as the following:

1. Occupancy patterns
2. Day-of-week occupancy patterns
3. Cancellation patterns
4. No-show patterns
5. Booking pace

When the demand goes down, the rates are reduced or lower rates are opened for the customers. When demand increases, the lower rates are closed and the effective rate goes up. Based on the way demand is increasing or the booking pace is changing, the rate changes. Interestingly, just the immediate profit is taken into account in this decision making, but the systemic impact of this rate change is not considered.

In a particular case it may result in profit, but the long-term impact on profit may or may not be positive. What needs to be done is to get sustainable profit gain and in

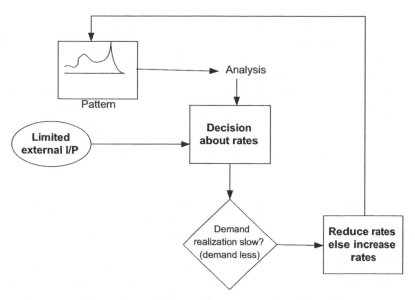

Figure 2.28   Hospitality industry decision making.

interest of all subsystems. Figure 2.28 depicts this learning and decision-making process.

In the case of some conference or function in the city, when rate across the city goes up, these types of systems impact on the selection of that destination in the future. Even the regular visitors of a city may choose another destination. Sometimes a particular chain or hotel may not see this impact, but there is systemic impact on overall tourism because of such decision-making schemes. These short-term pattern-based decision techniques may lead to some immediate revenue benefits, but in the long run these event- and pattern-based learning techniques may lead to disasters. There is need to consider the whole system and the long-term patterns over the period of time.

The observation clearly shows that the learning in this particular scenario is based on the objective function, which is maximizing profit. The learning is not based on target or desired system behavior. The decision-making issues may be due to either understanding and defining the desired system behavior or failing to learn based on desired system behavior.

2.9 SUMMARY

The need of an intelligent system to demonstrate more than what traditional machine intelligence has defined is growing day by day. Users have become more demanding. The typical limitations of learning are inability to build systemic view and inferencing based on limited information. Systemic learning is about understanding systemic impact of decisions and learning different interactions among the systems and subsystem to empower an decision-making algorithm to make systemic decisions.

Systemic machine learning needs to build holistic decision matrix for a particular decision scenario based on the available fragmented and partial information. Semi-constrained influence diagrams can be used to represent the particular perspective and the partial information. The representative influence diagram and decision matrix derived from it can help in multiperspective learning. System detections, mapping of system, and system analysis are required to understand system interactions and impact of the decisions on various parts of the system. Adaptive systemic learning allows the system to learn from the dynamic scenarios and build a decision matrix based on perspective analysis.

There are many aspects of systemic learning and analysis of information from different perspectives, and then integration of that information can help in building systemic scenarios. There is need to be incremental learning and making use of the additional information becoming available over the time.

Another important aspect of system learning is that it empowers a system by locating the highest leverage points in terms of time and decision space at the time of decision making. These highest leverage points allow the decisions and actions to be more effective. Actually the traditional learning is a special case of systemic learning where the system is identical to the decision-makers' views.

REFERENCES

1. Senge P. *The Fifth Discipline—The Art & Practice of The Learning Organization.* Currency Doubleday, New York, 1990.
2. Howard R, Matheson J. *Readings on the Principles and Applications of Decision Analysis*, Vol. 2. Strategic Decisions Group, Palo Alto, CA, 1981.
3. Wellman M. Fundamentals of qualitative probabilistic networks. *Artificial Intelligence*, 1990, **40**, 257–303.

CHAPTER 3

Reinforcement Learning

3.1 INTRODUCTION

In this chapter we will introduce the reinforcement-learning applications, fundamentals, and reinforcement learning from a broad and systemic learning perspective. An intelligent agent (IA) or any intelligent system performs actions based on the inputs received. There are many intelligent applications where historical information and learning based on historical patterns can exhibit required intelligent behavior. Unfortunately, this is not true for other classes of applications where the situation is dynamic and knowledge needs to be built continuously on top of learned facts. Furthermore, overall outcome of the decision is not based on a single decision or move. For example, while playing basketball, a basket results as a series of good moves. It is not just about the right or wrong move, but a series of good moves with reference to opposite players' positions can result in a basket. It is rather the goodness of the move in that scenario and its possible impact on the final outcome that helps in learning and decision making. Here, all decisions are moves that are not independent, and their mannerism is defined with reference to environment. The key aspects of these types of applications—whether it is basketball, football, or even some business process—are the role of the environment, measurement of goodness of rewards, and the feedback.

In all the above applications, there is a decision maker and environment. At any moment of time, the environment is in a certain state. In reinforcement learning, a learner or decision maker makes decisions and actions in association with the environment. Like an intelligent learning agent—an agent senses the environment and decides the best possible action with reference to the goal. The series of actions are taken in order to achieve the goal. For any action taken while solving the problem, an agent gets a reward or penalty from the environment. There is a set of trial-error runs. After such trial and error runs, the agent learns the best possible policy to solve the problem.

When we try to solve any problem, we get some outcome that can be used to measure the performance. There can be a number of actions taken to reach a goal. An autonomous agent senses and acts in the environment and chooses the optimal action to reach the goal through reinforcement learning. The agent should be able to

Reinforcement and Systemic Machine Learning for Decision Making, First Edition. Parag Kulkarni.
© 2012 by the Institute of Electrical and Electronics Engineers, Inc.
Published 2012 by John Wiley & Sons, Inc.

learn from all direct, indirect, or delayed rewards to choose the best set of actions. Various things and entities an agent interacts with include everything associated with an agent outside it. All these things together are called an environment.

Let us take an example of football where a sequence of actions of players results in a goal or a foul or penalty corner. The ultimate reward or the resulting value might be a win or a loss. But at every stage you get a reward for action. Suppose player "A" passes the ball to player "B." Player "B" takes the ball and kicks it in the direction of goal. This is a sort of positive reward. But in case player C of the opposite team reaches the ball before player A takes the ball and passes it to the player of their team close to the goal, the result is a negative reward or penalty. The percept sequence is input while the actions are performed based on percept sequences and the perceived state of the environment.

An agent is anything that is interacting with the environment through sensors and actuators. Typically, it perceives its environment through sensors. The actuators allow the agent to take actions with reference to environment or act upon environment through actuators. Any human being is an agent and is sensing the environment through his/her sensory organs such as ears, nose, skin, eyes, and tongue. He or she can act on the environment through hands, legs, and other parts of the body. An intelligent car can have sensors such as a camera, ultrasonic waves, and various other devices to measure distance, determine objects, and calculate light and weather conditions. It can have some mechanism to apply either gas or brake as an actuator to act on the environment based on perceived road and weather conditions. Actually an agent and the environment interact with each other continuously. For convenience, let us assume that each discrete time-step agent receives some representation of the environment's state.

An IA is an autonomous entity, which observes and acts upon an environment and directs its activity toward achieving goals. IAs may also learn or use knowledge to achieve their goals. They may be very simple or very complex: A reflex machine such as a thermostat is an IA, as is a human being and a community of human beings working together toward a goal. At any moment of time there are possible legal actions that an agent can perform. An agent's policy is nothing but implementing a mapping from states to probabilities of selecting each possible action. A typical relationship between an agent and environment is depicted in Figure 3.1.

Here percept refers to an agent's perpetual input at any instant in a given state. These inputs are received through sensors. These inputs actually build the view of the system, environment, or the world for the agent. Multiple sensors and multiple percepts can be obtained for the agent. In the case of a basketball game, the percept is to typically build a view about the positions of the players of one's team, position of players of the opponent team, time remaining, distance from basket, present score, location of ball, and so on. Since positions are changing continuously, there is a transition of state on account of change in percept. This may result because of movement by players, ball passing, whistle by referee, and so on. As the environment is dynamic and decision making needs continuous learning, static agents are of limited use. The concept of a learning agent, where the agent can keep learning continuously, is more suitable in the case of reinforcement learning.

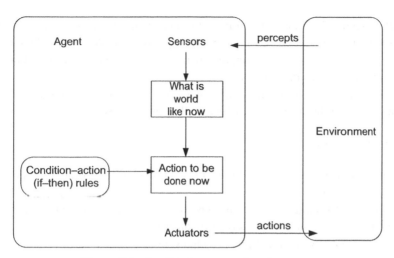

Figure 3.1   Intelligent agent and environment.

The learning agent can adapt to multiple and changing situations and can handle complex problems. It can succeed in a variety of environments. It has a learning element and a performing element. In short, a rational learning agent should possess the following properties:

1. It should be able to gather information—continuously or after a certain time interval, that is, periodically.
2. It should able to learn from experience.
3. It should have an ability to learn continuously.
4. It should augment knowledge.
5. It should possess autonomy.

Furthermore, there are many complexities with reference to the environment of the agent. The environment is generally dynamic and changing, and in real life the environment is not deterministic. One of the major limiting factors about available environment information is that it is not fully observable or rather it is partially observable. This makes it necessary for an agent to be flexible to perform intelligently and to use the partial information in the right context and effectively. The intelligence involves the inference about the unknown facts and taking the right actions in a partially known environment.

As we can see, intelligence demands flexibility. Flexibility equips the agent to negotiate with dynamic scenarios. To exhibit the required intelligence, we expect certain properties associated with flexibility from an IA.

By flexible, we mean that system should be able to adapt with the changing scenarios and should exhibit rational behavior in those changing conditions. For this purpose it needs to be

1. *Responsive*: Respond in a timely fashion to the perceived environment. It should be able to perceive changes appropriately and respond to the changes.
2. *Proactive*: Should exhibit opportunistic, goal-directed behavior and take the initiative where appropriate.
3. *Social*: Be able to interact (when they deem appropriate with other artificial agents) with humans in order to complete problem solving.

Other properties an IA should have are

1. *Mobility*: It is recommended that it should be mobile. It should not get just a static percept.
2. *Veracity*: IA should be truthful. The true correct and rational picture of the environment should be perceived by an agent.
3. *Benevolence*: Avoid conflict—do what is told.
4. *Rationality*: It should exhibit rational behavior. It is more like logical behavior.
5. *Learning*: It should learn from changing scenarios, state transitions, and behavioral changes.

As discussed in the previous chapter, intelligent systems need to have learning capability. Learning with reference to what is already learned, along with learning based on exploration in case of new scenarios, is required. The agent, in order to deal with dynamic scenarios, needs to handle both exploration and exploitation. There is a need for adaptive control and learning abilities. Before discussion about adaptive control, let us discuss about the learning agent.

3.2   LEARNING AGENTS

An agent needs to operate in individually unknown or rather changing environments many times. The already provided knowledge may not be enough to deal with new and changing scenarios. Also, an already built knowledge base does not allow an agent to operate in an unknown or new situation scenario. This makes it necessary for agents to have the ability to learn and negotiate with new and changing environments. This ability to learn can allow agents to respond to new or unknown scenarios in a logical way. Furthermore, learning can help in making improvement in behavior as it proceeds or comes across more and more scenarios. Most importantly, it will allow agents to learn from experience. There are three important elements in a learning agent:

1. Performance element
2. Evaluation element
3. Learning element

The learning element is responsible for making improvements, while the performance element is responsible for selecting external actions. Here the performance element is an agent without the learning element. The evaluation element tries to coordinate among selected external actions to build a platform for learning. In short, any simple or complex agent with learning elements forms a learning agent. The learning element needs feedback based on measurement of performance—that is, how the agent is doing. This feedback element drives the learning actions. A critic provides this feedback required for learning. The critic percepts agents' success and provides feedback. Based on the design of a performance element, many designs for a learning agent are possible.

Another important part is the problem generator, which is responsible for suggesting actions that will lead to new and informative experiences. Feedback may come in the form of penalty or reward. This penalty or reward helps in improving the performance of agents and building knowledge base. The penalty or reward is based on observations of behaviors which are the result of actions (Figure 3.2).

Though the learning is possible in different ways, the concept of reinforcement learning tries to handle the issues of exploration and exploitation and improvement through trial-based learning to negotiate with this type of scenario. In this chapter, we will discuss the aspects of reinforcement learning in context of this issue. The desired control strategy chooses a sequence of actions from the initial state, which maximizes rewards.

In a real-life scenario, learning with a teacher, which is more of the form of supervised learning, is not always possible. Agents can use a predictive model of environment—such as what the situation looks like, what it would be after a specific action, and what is the response of their opponent. Some random moves with simple background logic are explored. While exploring with these actions, an agent needs to know the overall impact—this comes in the form of reward or reinforcement (Figure 3.3).

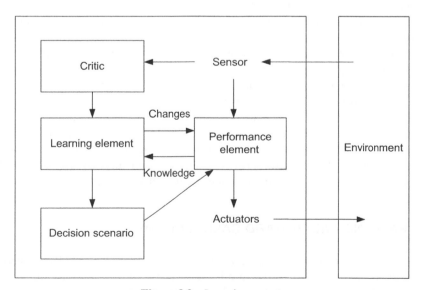

Figure 3.2   Learning agent.

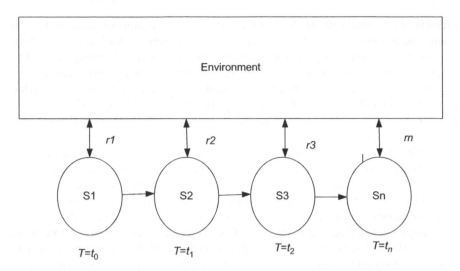

Figure 3.3   How rewards come in reinforcement learning.

Rewards always come at the end of the game—there can be rewards coming more frequently in games such as basketball, boxing, etc. An input percept sequence is used to understand the environment—reward comes as a part of percept. The mechanism needs to be in place for an agent coupling with the environment to understand rewards in a timely fashion. There is an advantage in learning when we can use these rewards coming quite frequently. The optimal policy needs to be chosen based on the percept of sequence and rewards based on that and—the optimal policy is one that maximizes the total expected rewards. The objective of reinforcement learning is to understand the observed rewards and determine the optimal policy that maximizes cumulative rewards.

As we discussed above, reinforcement learning is about coming up with the specifications and guidelines about how an agent should change its policy as a result of experience through explorations. One important aspect of reinforcement learning is about understanding the boundary between an agent and environment. This boundary is generally defined based on the area that an agent can control arbitrarily. The area beyond the control of the agent is considered outside oneself—that is, the environment. Some area in the environment is known to the agent while some may not be known to the agent. The reward source is generally placed outside the agent. Still the agent can define internal rewards or sequence of internal rewards.

3.3   RETURNS AND REWARD CALCULATIONS

The goal of an agent is to maximize the reward it receives in the long run. The selection of actions to maximize expected return is the objective of learning. The cumulative rewards can represent the returns. These rewards are the ones obtained at time intervals. Let T be the time from where learning began, while t is the present time stamp. Hence total rewards R_T is given by the following equation. Hence:

$$R_T = r_{t-1} + r_{t-2} + r_{t-3} + \cdots + r_T$$

The agent–environment interaction breaks into what we call episodes. Each of the episodes ends with a special state called a terminal state. It is assumed that the agent–environment interaction is divided into a number of identifiable episodes. The corresponding tasks are referred to as episodic tasks. But practically it is not always possible to divide this interval into a number of such distinguishable episodes, especially in the case of continual process control where there are continuing tasks. Another important aspect is that the total reward is represented as a sum of discounted rewards an agent receives over the future.

3.3.1 Episodic and Continuing Task

To be precise in the case of episodic tasks, we consider a series of episodes, each of which consists of a finite sequence of time steps. Though the consideration is generally confined to a single episode, it makes sense to consider it in association with rewards during adjacent episodes.

The notions of episodic tasks and continuing tasks have their own limitations. As we proceed, we will try to get the better of both worlds. Figure 3.4 depicts a typical state diagram.

The rewards in an episodic task can be represented by

$$R_t = \sum_{k=0}^{T} \gamma^k r_{t+k+1}$$

3.4 REINFORCEMENT LEARNING AND ADAPTIVE CONTROL

Reinforcement learning (RL) is more like trial-and-error-based learning in association with the environment. All control problems need to manipulate dynamic system inputs such that behavior meets the required specifications. A measure of the aggregate future rewards used in RL is the so-called value function. It represents the main objective function for the RL problem. Typically, at each time instance the value function is estimated and the action that maximizes the value function is taken. The value function represents the total value created rather than that associated with a

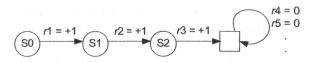

Figure 3.4 State diagram.

single percept. In what follows, we will have a very brief review of the problem of value function estimation. Rewards are the results of certain actions taken by an agent with reference to the environment. The reward calculations are external to an agent and cannot be internal to it. This is because the rewards are with reference to environment and are not in control of an agent—only with systematic learning can an agent work on improving itself. This is very clear from the following example—if there is a series of right moves, the player can win the match—but this reward comes externally and is not in complete control of the player. In short, the agent–environment boundary represents the limit of the agent's absolute control. It is one of the trickiest parts to determine the agent–environment boundary. It is generally determined with reference to a particular state, actions, and rewards. Even these boundaries are sensitive to decision scenarios. As we move ahead in this book, we will discuss this part in greater detail. We can even look at reinforcement learning as an abstraction provided with goal-directed learning from the various interactions. Any problem where learning is required with goal-directed behavior is represented in the form of agent, environment, and interactions between them in the form of actions, rewards, and states. The choices made by agents are represented as actions, while the states are the basis for making choices.

The goal of an agent is to maximize the total rewards it receives. The concept of adaptive critics are the critics that give feedback in an uncertain environment. It is also the name of algorithms that approximate dynamic programming (DP).

To adapt means "to change (oneself) so that one's behavior will conform to new or changed circumstances." The reinforcement learning is trying to achieve the same objective. An adaptive controller is formed by combining an online parameter estimator, which provides estimates of unknown parameters at each instant, with a control law that is motivated from the known parameter case. The way in which the parameter estimator (also called as adaptive law in literature) is combined with the control law gives rise to two different approaches. In the first approach, referred to as indirect adaptive control, the plant parameters are estimated online and used to calculate the controller parameters. In real life, the environment is changing and a simple rule based on conditional controls cannot cope with the dynamic environment. For adaptive control there is the need for three components:

- Environment sensor
- Reference model
- Controller with adaptive function (Figure 3.5)

For adaptive control the system needs to sense the environment and response from the environment continuously. Based on rewards with reference to a sequence of actions, the learning results on adapting to a new environment scenario. A simplified model of adaptive controller with reference to reinforcement learning is depicted in Figure 3.6.

In real-life problems, AI or any intelligent system will likely not negotiate against static and simple environments or scripts but rather against dynamic (changing and

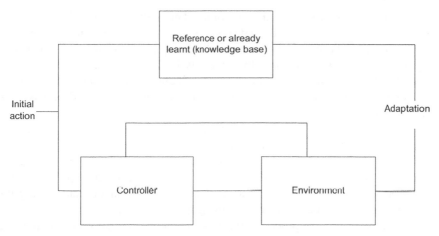

Figure 3.5   Reinforcement learning—adaptive.

complex) environments and even in some cases against intelligent humans and changing tactics. Humans will employ more challenging and changing tactics (that are never seen before) and hence the overall environment response may not be fully predictable based on previous experience. In a game type of environment, opponents may use one tactic until it seems to fail and then switch to a new, distinct tactic or keep switching their tactics between each game. In some cases, opponents may even keep

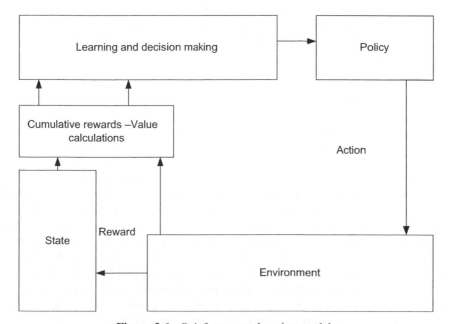

Figure 3.6   Reinforcement learning model.

switching the tactics in spite of the outcomes. This will cause traditional learning systems to fail using their just-learned or simple fuzzy rule selection. It is therefore interesting to see if the system can, instead of learning a static tactic, adapt to distinct changing scenarios and responses to tactics itself. There is a need for switching between fixed, distinct strategies as a response to change in environment. Intelligence here cannot be based on mere rules or known facts but can instead be based on dynamic strategic response.

There is active as well as passive reinforcement learning. Passive reinforcement learning has a fixed policy while in case of active reinforcement learning an active agent must decide what action to take. The agent should understand the connection among states or how the states are connected. In adaptive dynamic programming, agents work by learning the transition model of the environment and solving the corresponding Markov decision process using a DP method.

In deterministic control processes there is an assumption that state variables are identifiable and observable. The assumption here is further extended and states that possible decisions are known and there is full knowledge about the cause-and-effect relationship. This actually is not true in many real-life scenarios. The world has the following parts:

1. System
2. Environment
3. Agent

An agent is part of a system and also interacts quite frequently with the environment. Figure 3.7 depicts a typical learning framework. Here an agent has a learning system, sensors, and a decision system. The agent and its learning system interact with the environment on a continuous basis.

3.5 DYNAMIC SYSTEMS

A dynamic system is one that shows the time dependence in ambient space. In the case of decision making for a dynamic system, there is need of learning based on exploration. In this section we will discuss the role of reinforcement learning in the case of dynamic systems. We will discuss in detail the active reinforcement learning. Active reinforcement learning is associated with an active agent. An active agent is an agent with the capability to make a decision on what action to take. Greedy agents fail in learning an optimal policy and real utility value for other states and especially for the dynamic system. In the absence of information or without a model of real environment, the optimal selection of a state leads to suboptimal results. In real problems, the agent is not aware of the true environment and hence computing an optimal action is not possible. Thus, there is need for exploration to maximize the rewards. In the case of dynamic systems, novel information along with the existing information should be used effectively; novel information becomes available with changing scenarios.

Environment

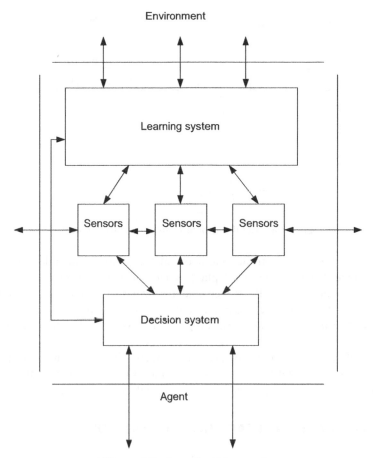

Figure 3.7   Learning framework.

3.5.1 Discrete Event Dynamic System

Before building a learning infrastructure for dynamic systems, there is a need to understand a dynamic system and its behavior in greater detail. Also, it is important to understand what type of learning opportunities a dynamic system creates and expected learning behavior for dynamic systems. Discrete event dynamic systems (DEDS) are the asynchronous systems where occurrences of discrete events in the system trigger the state transitions. DEDS can be represented by the quadruple:

$$G = (X, \Sigma, U, \Gamma)$$

Here X is finite set of states, Σ a finite set of events, U a set of admissible control inputs, and Γ a set of observable events, and this is a subset of Σ. Event-driven systems can be modeled using DEDS. Bellman's DP algorithm can be used as a mathematical foundation for optimal control of systems. An adaptive control dynamic system has an

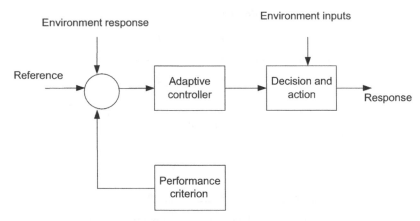

Figure 3.8   Reinforcement learning adaptive control.

adaptive controller. This controller interacts with the environment or decision region. The environment, which may be a plant or part of a system, gives some sort of response for every action. Reference and performance criteria are input for an adaptive controller. This typical adaptive controller needs to adapt control mechanism based on performance and outcomes. A typical adaptive control system is depicted in Figure 3.8.

Here a reference represents already learned facts while environment response is based on exploration. A DESD is called an observable one if the sequence of observable events or sequence of percepts can be used to exactly determine the current state.

3.6   REINFORCEMENT LEARNING AND CONTROL

Reinforcement learning and mathematical-optimization techniques work very closely. Formation and optimal use of a value function is the key concept in reinforcement learning. Reinforcement learning learns the value function while interacting with the environment. This value function can be directly used to implement a policy. The agent needs to play an important part in making reinforcement learning feasible where it should perform activities like storing and manipulating the value function. This is typically done by evaluating a policy and working on policy improvement. Furthermore, agents must provide the appropriate control action for a given state. So the two most important functions of agents in reinforcement learning are

- As a reinforcement learner
- As a controller

3.7   MARKOV PROPERTY AND MARKOV DECISION PROCESS

As we discussed in pervious sections, a decision is made based on the state of the environment. We will discuss Q-learning and various aspects of Q-learning with reference to control policies in the next section. In this section we will discuss Markov Property and decision process and value function.

The state of both the environment and system can impact decisions and outcomes. The state in any scenario is represented by whatever information is available with the agent. A state signal that can retain all relevant information (here state signal summarizes past sensations completely) is called a Markov or one that possesses a Markov property. In short, it is not about a sequence of states but is, instead, about the ability of the present state and retaining or summarizing ability in the present state to determine the future. It is independent of the path or sequence in the past. If the environment response at state corresponding to time $t + 1$ only depends on the state and action representation at time t, then the state signal has the Markov property.

$$P(X_t \in A|F_s) = P(X_t \in A|\sigma(X_s))$$

If the state signal has the Markov property, on the other hand, then the environment's response at any point of time depends only on the state and action representations at that state, in which case the environment's dynamics can be defined by specifying only

$$P(X_n = x_n|X_{n-1} - x_{n-1} \ldots X_0 = x_0) = P(X_n = x_n|X_{n-1} = x_{n-1})$$

In this case, the environment and task as a whole are also said to have the Markov property.

Decisions and values in a Markov property are the functions of only the current state, and hence it is very important in the case of reinforcement learning. A decision-making process and reinforcement-learning task that satisfies a Markov property is called a Markov decision process (MDP). If the state and action spaces are finite, then the decision-making process is called a finite MDP.

3.8 VALUE FUNCTIONS

Value functions are the functions of a state, which determine how good the state is and how beneficial a particular action is in the said state. How good comes from the notion of expected future rewards. How good is generally decided based on the future rewards or expected returns.

$$V^\pi(s) = E_\pi(R_t|s_t = s) = E_\pi\left(\sum_{k=0}^{\infty} \gamma^k r_{t+k+1}|s_t = s\right)$$

The value of taking action a in state s under policy π is given by action-value function for policy Π-$Q^\Pi(s, a)$. The action-value function is given below:

$$Q^\pi(s, a) = E_\pi(R_t|s_t = s, a_t = a) = E_\pi\left(\sum_{k=0}^{\infty} \gamma^k r_{t+k+1}|s_t = s, a_t = a\right)$$

A reinforcement-learning task is about finding a policy that maximizes long-term rewards. There is always one policy that is always better than or equal to all other policies, and this policy is referred to as an optimal policy. There may be more than one optimal policy. The optimal policy is denoted by Π^*. These policies share the same state-value functions called optimal state-value functions and are denoted by V^*. Optimal policies that share optimal state-value functions are called optimal value functions and are denoted by Q^*.

3.8.1 Action and Value

The learning results in the selection of an action or sequence of actions. The action-selection decision is done based on the value of an action. The true value of any action is the mean reward received when that action is selected. One simple way is deciding the value by averaging the actual rewards received when the action was selected.

$$Q_t(a) = \frac{(r_1 + r_2 + \cdots + r_n)}{n}$$

For small values of n, these values may vary; but as n starts increasing, the Q value converges to the actual value of the action denoted by $Q^*(a)$.

Apart from this simple method, there can be various different ways to estimate the value, and those methods that converge quickly to actual value and more accurately are preferred.

3.9 LEARNING AN OPTIMAL POLICY (MODEL-BASED AND MODEL-FREE METHODS)

In the previous section, we discussed methods for obtaining an optimal policy for an MDP assuming that we already had a model. The model represents the knowledge about state transition that can be represented in mathematical form. Knowing this model in advance and making its effective use is one of the goals of reinforcement learning. There can be direct adaptive controls or indirect adaptive controls. There are two possible policies:

- *Model-Free Policies*: Here a controller learns without learning the model.
- *Model-Based Policies*: Here a model is learned and used to derive a controller.

The basic problem with reinforcement learning is determining whether the recently selected and taken action is good or bad. One strategy that is discussed so far is waiting until the end results: It rewards if the outcome is a good one, and it gives a penalty if the outcome is a bad one. Temporal-difference methods proposed by Sutton are about using insight from value iteration to adjust the estimated value of the state based on immediate rewards and the estimated value of the next state [1]. We will discuss temporal-difference learning strategies in the next section.

3.10 DYNAMIC PROGRAMMING

The focus of dynamic programming (DP) is more on improving the computational efficiency by dividing the problem into subproblems. DP tries to solve the problem in stages. It is the collection of algorithms that is used to decide optimal policies through computation, provided there is a perfect model of environment available—it might be in the form of MDP. The utility of DP algorithms in case of reinforcement learning is limited due to the need of the perfect model of the environment. Still we would like to have a brief introduction to DP in this chapter, as we may need to refer to it in subsequent chapters. Prior to that, it is important to understand what a dynamic system is and what a partially observable dynamic system is. DP can be applied in the case of discrete as well as continuous time. Dealing with inconsistency and leading to an optimal solution are the objectives of DP.

3.10.1 Properties of Dynamic Systems

Change is at the heart of the dynamic systems. Mathematically, though, the systems are in a particular state at any point of time and can be represented by set of real numbers. For the dynamic systems we are referring to here, only a limited view is available. With time the view changes. In short, the impact of decision and actions is time dependent.

$$T = t_0 \ldots T = t_n$$

Here the impact is observed for duration T.

Figure 3.9 depicts the concept of a dynamic system with reference to time. There are continuous changes, and each change and event are used by the agent for learning. Furthermore, even the action taken by the agent also results in event or change in scenario.

3.11 ADAPTIVE DYNAMIC PROGRAMMING

Adaptive DP combines the concepts of DP and reinforcement learning. In this case, the adaptive critic provides the feedback in the form of rewards or penalties. The adaptive critic is always keeping track of the desired performance. It supplies the method for optimal control. Figure 3.10 depicts a typical structure of an adaptive DP-based learning framework.

Adaptive value calculations are based on its interaction with the environment by critics. The policies are updated based on critic feedback. The adaptive-critic learning allows us to deal with a dynamic environment. As we can see, adaptive DP combines concepts of DP and reinforcement learning.

3.11.1 Temporal Difference (TD) Learning

DP can be used to solve the learning problem and to determine the optimal policy. DP and similar methods for determining an optimal policy are costly, and we seldom have

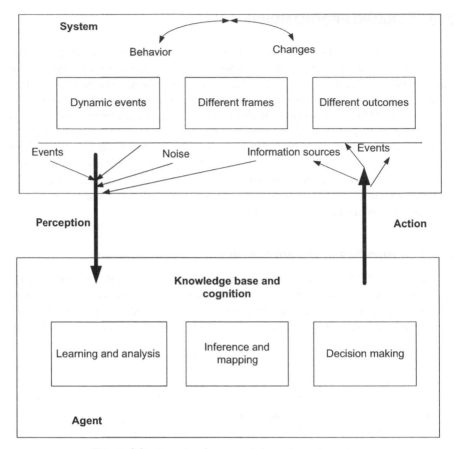

Figure 3.9 Learning framework for a dynamic system.

the knowledge about the complete environment. TD learning is the combination of Monte Carlo simulation ideas and DP ideas. TD methods can learn from direct experience like Monte Carlo methods without modeling the environment and, like DP methods, update the estimates based on other learned estimates—without waiting for the final outcome.

Monte Carlo methods need to wait until the end of episode while TD methods need to wait only until the next time step. After every time step—that is, from t to $t + 1$—it immediately forms a target and makes the useful update using the observed reward r_{t+1} and estimate value $V(s_{t+1})$.

The simplest TD method is known as TD(0), and it can be represented as follows:

$$V^{\Pi} \leftarrow V(s_t) + \alpha[r_{t+1} + \gamma V(s_{t+1}) - V(s_t)]$$

A simple TD method is based on just one next reward, while the Monte Carlo method is based on the entire sequence of observed rewards from that state until the end of the episode. Both methods have pros and cons. One intermediate method is to

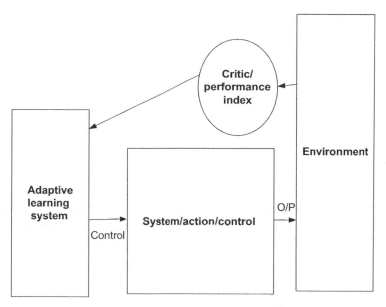

Figure 3.10   Adaptive dynamic programming-based learning.

use intermediate number of rewards. The Monte Carlo target is an estimate—here the expected returns are not known.

In intermediate reward methods there are methods based on the number of steps in backups that are been used. This includes one-step backup, two-step backup, three-step backup, and n-step backup. An n-step backup is still the TD method because it changes earlier estimate based on how it differed from the later state. The only difference is that instead of having just one step, it happens n steps later. It is called the n-step TD method.

Example: While working on a software project every day, one would like to predict how long it will take to complete the project. One of the simple ways is the feedback coming at the end of the project deadline. There can even be some sort of feedback after every episode, such as feedback after development of modules. It can be further extended so that there can be feedback after every day's development and based on everyday's project build. The value of expected man-hours required to complete the project is the value. Typical burning calculations in case of a scrum methodology can be used for TD learning. Predicted project-completion time is calculated again at the end of every episode.

3.11.1.1   *Advantage of TD Predictions*   There are many advantages of a TD method over the Monte Carlo method. One of the most important advantages is that it does not need the model of environment of its rewards—and also next-step probability distribution. Another important advantage, which is more significant in real-time systems, is that it can be implemented in a fully incremental fashion and online. Hence it can be very useful for dynamic scenarios. TD methods need to wait only one

step at a time. This property is very useful in case of many real-life problems where episodes are very long. Learning based on each transition even helps in speedy learning. Hence TD methods generally converge faster than Monte Carlo ones.

3.11.2 *Q*-Learning

Q-learning does not estimate a system model. *Q*-learning is based on estimating the real-valued function Q of the state and actions, where $Q(x, a)$ is the expected sum of discounted reward for performing action a in state x and then going and performing optimally thereafter. A *Q*-learning agent learns as a function that is more like action-value function as discussed in the previous section. This action-value function is also referred to as a Q function that gives the utility of taking a particular action in a given state. *Q*-learning is a technique that works by learning an action-value function. The environment is dynamic, and it is important to understand how an agent can learn an optimal policy for an arbitrary environment. The properties of *Q*-learning algorithm are given below:

- It is a reinforcement algorithm that is incremental—incremental in the sense that the weights are changed after each transition.
- This one is direct.
- It guarantees convergence in case of discrete case and finite number of states.
- It can learn from any sequence of experiences.

Q-learning does not store actions and values, but it stores Q values. For a given state, say s and action a, the optimal Q value is represented as $Q^*(s, a)$. Here $Q^*(s, a)$ is the expected total reinforcement that is received by starting in state s—performing action a on the first step. After that an optimal action is performed. The value of the state is the maximum Q value of that state. So obviously the action associated with this maximum Q value—hereafter referred as Q value of the state is the policy of the state.
Let P^* be an optimal policy.
The quality of state action combination is represented by

$$Q : S \times A \rightarrow R$$

$$Q(x, a) = E\{r_k + \gamma \max_b Q(x_{k+1}, b) | x_k = x, a_k = a\}$$

Q-learning maintains an estimated Q value. The Q function combines information about state transitions and estimates future reward without relying on state-transition probabilities. The less computation and independence from the explicit state-transition probabilities make *Q*-learning more attractive.

3.11.3 Unified View

All of the reinforcement learning methods discussed so far have many things in common.

1. The purpose of using them is to estimate value functions. Value functions are the key concept in reinforcement learning and they decide the overall learning track.

2. The methods discussed so far operate by backing up values along actual or possible state trajectories. It is based on actual or estimated rewards.

3. These methods follow the strategy of generalized policy iteration (GPI), meaning that they maintain an approximate value function and an approximate policy, and they continually try to improve each on the basis of the other.

As a unified view these value functions, backups, and GPI play a role in modeling intelligence and hence deciding the learning track. Finally, the purpose of reinforcement learning is to produce a uniform view based on available learned knowledge and the response from the environment to produce learning pointers.

3.12 EXAMPLE: REINFORCEMENT LEARNING FOR BOXING TRAINER

Reinforcement learning can be used in case of an automated boxing trainer. The opponent may make new moves, and depending only on the past information or knowledge may not work.

The boxer has some internal state, along with perception of the external world, and there is a need to understand the intention of the opponent. The boxer has a perception system capable of giving him the relative angle and distance to the other boxer; an inference mechanism will tell the possible intention, probable move, and impact of each move. Now based on these actions the trainer will learn about the capabilities of the opponent and can respond accordingly to train the opponent for the competition.

The learning system decides intentions based on reward and punishment scheme used. While this is happening, a trainer may move slightly away from the boxer. In normal conditions, it observes some characteristics of the opponent.

There are many applications where the behavioral pattern of the system is dynamic and the knowledge base needs to be improved on every exploration. Reinforcement learning offers the ability of dynamic learning to deal with such applications.

3.13 SUMMARY

Learning does not happen in isolation. Learning generally results through interactions and responses. Reinforcement learning tries to overcome some of the limitations of traditional learning. Reinforcement learning uses exploitation of already learned facts and exploration based on new actions and scenarios. It tries to learn with reference to the environment, and learning takes place in association with the environment.

Reinforcement learning tries to use the most important part of learning; that is, it generally results in association with the environment. The exploration activities try to sense the response of environment for action. For every action performed we get a certain reward. These rewards reflect the utility and relevance of the action. These rewards build guidelines for learning. During exploration the knowledge is built based on the response the agent gets from the environment. Reinforcement machine learning builds the foundation for next-generation machine learning where learning is no longer an isolated activity. As we go through the next few chapters, these ideas are extended to cooperative and systemic learning to make holistic learning possible.

REFERENCE

1. Sutton R. Learning to predict by the method of temporal differences. *Machine Learning*, 1988, **3**(1), 9–44.

CHAPTER 4

Systemic Machine Learning and Model

4.1 INTRODUCTION

We have studied reinforcement learning in the last chapter. In this chapter, we will elaborate decision and learning models using the systemic perspective. The most important aspect of a systemic model is developing the understanding of the system, its parameters, and decision boundaries. Learning is necessary to provide the right decision, and it is very closely bound to decision scenario. Learning cannot be absolute, and it evolves with changes in environment. Every decision scenario needs different parameters, and the decisions need to be context specific. It is important to understand what we expect from a particular decision.

The learning is a cooperative process. This cooperation needs integration of data coming from different sources, interpretation of information from different perspectives, and inference with reference to different contexts. Systemic machine learning models determine the system, make decisions about parameters, and with reference to context provide the best possible information. In this model, the decision making is not just about action and outcome, but it is about a series of actions and a series of outcomes in time, decision, and system space. The learning in case of dynamic scenarios needs to determine these parameters on an ongoing basis. The model should select the best possible set of parameters, choose the optimal boundaries of the system with reference to the decision scenario, and deduce the decision context. The strength of learning depends on the accuracy in determining decision context. This can be done in a more semisupervised way with use of inference mechanisms to deduce unknown facts and build system view.

The data inputs come from various sources. These inputs come from systems and subsystems and are in the form of data, behaviors, short-term results, and decision outcome and even in the form of patterns. These inputs or rather data coming from different sources build a platform for the context determination. Systemic models are contextual, time-based, and based on inference for long-term outcome. In this chapter,

Reinforcement and Systemic Machine Learning for Decision Making, First Edition. Parag Kulkarni.
© 2012 by the Institute of Electrical and Electronics Engineers, Inc.
Published 2012 by John Wiley & Sons, Inc.

we will discuss a systemic learning (SL) and decision-making framework. The models for learning are data driven and hence always try to determine the likelihood for fitting. The models are based on the philosophy of continuous learning and learning with reference to environment. Here environment is determined with reference to decision scenario and unlike the environment we referred to in the previous chapter. The sources of information, integration of multisensor data and context-based decision making are some of the important factors of the systemic decision making.

Knowledge capture, knowledge building, and knowledge use are some important aspects of this overall learning process. The knowledge needs to be systemic in nature. The difference is subtle between building and gathering. The systemic knowledge building takes place continuously with reference to the system and is not built in isolation. This chapter also tries to look into decision-impact analysis for the sake of continuous learning. For any action there are many possible outcomes that can be seen—some are immediate and visible, some are minute but have greater impact in some other context, some may come much later, and some are beyond visible frame, while others are difficult to map to action. Any action results in many outcomes, and it is important to know the relevance of those outcomes in a particular decision context.

Learning is based on the feedback the learning engine receives from the system behavior. Getting the right feedback and interpreting it to build the right context for the learning are some other aspects that are dealt with in this chapter. In this chapter, we will study frameworks and models that can help in building a system view and provide the best possible decision for a given decision context.

4.2   A FRAMEWORK FOR SYSTEMIC LEARNING

An SL framework refers to a skeleton structure designed to support systemic machine learning and decision making. The decision models are based on choices available for decision making. Rational decision-making models typically decompose the elements of a decision problem. This helps in getting insights into choices, uncertainties, and outcomes. Descriptive models are more heuristic based, and decision making is based on how the things actually work. The next level is about understanding the situation. This we will refer to as "situation awareness." Situation awareness is awareness about actions, system, and relationships. It provides the primary basis for subsequent decision making and performance in the operation of a dynamic system. This situation awareness develops perspective. It requires a mechanism to quickly collect data from the environment and then detect, integrate, and interpret this data. This actually results in building the systemic knowledge.

Situation awareness involves detecting relevant elements and perceiving the status, attributes, and dynamics of relevant elements in the environment and with respect to environment. The car driver needs to perceive information about his car—speed of the car, road condition, type of road, speed of cars in adjacent lanes, directions, milestones, etc. The comprehension of the current state is another important part. This includes understanding of significance of each of the parameters in light of decision scenario and systemic goal. Based on the comprehension of the significance of the information, the overall context is built. The overall system view is inferred, and based on that the impact of each of the probable action is determined.

The perspective of the elements in problem space with reference to decision space and time along with comprehension of the meaning in that context need to be taken into account for systemic decision making. Nonsystemic models tend to abstract rationality and neglect systemic complexities. Context and "situation awareness" have a few things in common.

Systemic learning framework introduced in this section deals with six important aspects of learning:

1. *Tapping the Information*: It is the process of identifying information-relevant sources and tapping information coming through these sources and outcomes. Changes and patterns are sensed. The information tapping needs to be context based. The techniques such as context-based data mining can be used for getting context-based data.

2. *Knowledge Building*: Knowledge building includes the use of mined data, integration of information, and mapping the information with reference application. Integration of possibilities, feasibilities, impact, and efficiency based on inference is used for knowledge building. This knowledge base creates a platform for decision making and learning. There are two levels of learning—data and pattern-based learning and context-based learning. This knowledge base is focused on data-based learning.

3. *Analyzing Decision Scenario*: A decision scenario is a rather specific case where we need to apply and use learning to get effective decision or outcome. The decision scenario is analyzed from systemic perspective. A decision scenario is a representation of decision state, decision objective, and parameters needed for decision making. The clear understanding of a decision scenario can help in learning for the decision scenario. A decision scenario helps in understanding the essential factor and giving the right weight to a parameter.

4. *Decision About System Boundaries*: System boundaries define the region in which a decision is made and will impact. These boundaries define the effective area of the impact of decision. System analysis and decision scenario are used for detecting system boundaries. These boundaries are not generic and are confined to decision scenario.

5. *Context Building*: The information is generally available in bits and pieces. The overall decision context for the particular decision scenario is the key for decision making. Boundary detection and system parameters are used for building context.

6. *Action Space and Impact Space Analysis*: The action space and impact space are different, and hence analyzing the space and learning with reference to action space and decision space is required. Based on context, the action space and decision space are determined—context along with decision space and action space are used for SL and decision making.

Here action space is the area in which an intelligent agent moves and a "decision system" makes decisions to seek the desired outcome—for example, an auto closing door: action space is the door and its framework. The impact space can be much beyond the action space—for example, if the door opens in the case of a wrong swipe card; that is, a wrong decision may impact the whole shop. It may sometimes impact areas beyond the shop and even that result may not be visible right away. If the door remains open for a longer duration, it may impact on air conditioner and its compressor.

Decision space is the range of effective and valid choices available with the decision maker. Decision space defines the region and the points along with the choices available and guides about the scope of the decision making.

Complex decision problems have many parameters. In some cases, the relationships among these parameters is not clear. The relationships reveal at various stages during the process of solving. These problems need analysis and a clear understanding of the relationship to lead to a proper solution. In many decision problems it is perceived that the action space and impact spaces are the same. Ideally action space and decision space are not the same in most of the cases. As discussed above, action space refers to space where a decision-making solution moves, makes decisions, and takes actions. Before going into details of the framework, let us define impact space with reference to systemic machine learning.

4.2.1 Impact Space

Impact space refers to the region or space where the impact of actions taken in action space can be experienced. This includes direct as well as indirect impact. Since the whole world is connected, impact space can be whole world; but for the purpose of convenience and practical application building, impact space is defined as the space where the impact of the actions is felt and is measured above the threshold defined by systemic decision-making solution.

Figure 4.1 depicts an SL model. The information is gathered from all available information sources. This information is imperfect information as it may be partial, may contain noise, and may also contain information that may not be relevant from the decision scenario. This is raw information acquisition. This information along with systemic inputs, patterns, feasibilities, impact, and possibilities is used to build knowledge. This knowledge is richer and contains mapping among information artifacts with reference to patterns and feasibilities. This knowledge and decision analysis helps to build the comprehensive decision scenario. Decision scenario and system information helps us in boundary detection. The system boundaries help us to refine the knowledge and can help us in making the right decisions. With the entire available information, context for decision making is formed and this context is used for systemic machine learning and decision making. Action space and decision space along with contextual information makes systemic machine learning possible.

Figure 4.2 depicts the typical information-integration scenario for systemic decision making. Here information comes from various subsystems—rather, from the sources from different subsystems.

With reference to the above discussion, the purpose of an SL framework is to understand a system and provide decisions and learning optimal for the whole system

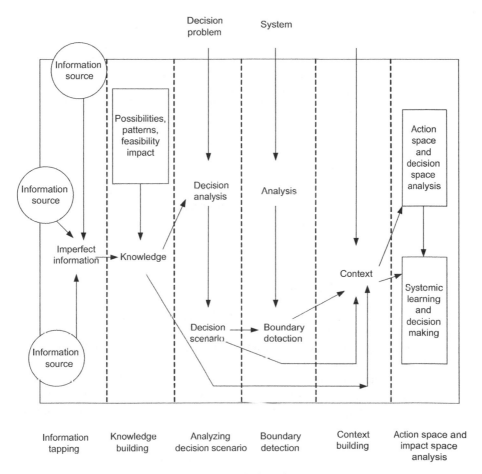

Figure 4.1 Systemic learning model.

with reference to a decision scenario. The too much generalization of a system makes the overall computation and dependency determination very complex. A simplified system representation is depicted in Figure 4.3.

Multiple information sources may bring heterogeneous information. This information may be incomplete and full of noise. The information from one source may complement information from other sources. Cooperative machine learning can be used to learn effectively from more than one source of information. In cooperative machine learning, different information sources and agents interact with each other. The conclusion or overall learning in this case does not follow directly from the data coming from one source or mere integration of data coming from different sources. The challenge lies in selecting the right data and a cooperative learning framework that can allow learning from data coming from multiple sources.

A typical cooperative learning system is depicted in Figure 4.4. Here communication media offers a platform for different information sources to interact with one

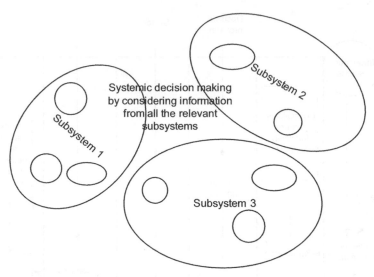

Figure 4.2 Information integration.

another. These information sources are generally intelligent agents or data reposito-ries. Evaluation of learning, decision parameters, inputs from critic along with individual learning is used by the learning controller. The cooperative learning gets feedback from environment in case of any exploratory action.

Here the interactions among outcomes come through multiple agents, or rather multiple sources are used for learning. Learning results from the cooperation between multiple agents or interpretations of those outcomes. It is about understanding multiple

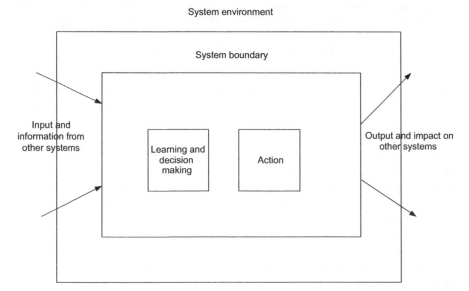

Figure 4.3 Boundary and environment of the system.

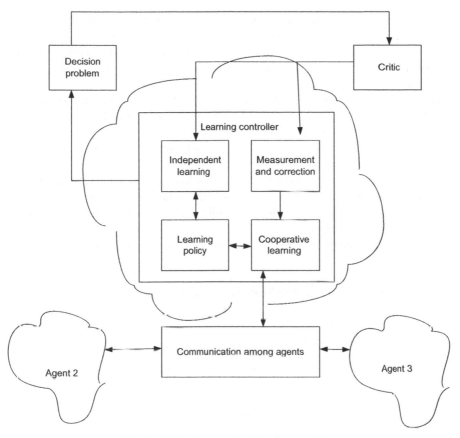

Figure 4.4   Cooperative learning model.

actions and resulting multiple responses in the system and decision space. This learning at first level results from the mapping between actions and changes in systemic behavior. Since the system is generally a much broader concept than decision space, cooperative and concurrent machine learning is used. The challenge remains to adapt the context and to come up with an overall context for decision making. There are a number of separate smaller decision and action spaces. Though the decision space for these actions might be different, impact space for these different action spaces can be identical. The impact space for different and smaller but relevant decision spaces might be much larger and can be identical. A learner explores its impact space while doing these actions. In cooperative learning, the overall learning changes with multiple learners and the overall learning builds the context while learners change their behavior with the development of context and clarity in systemic aspects.

In approximately all of the learning systems, agents learn individually. But the main favorable attribute of real-time systems (which are generally multiagents systems) is the ability to learn from the experiences of multiple agents or different information sources. Learning from the cooperation of information among the multiple information sources is the key. Moreover, agents can consult with more expert agents or get advice from them. This allows building more knowledge. Improvement in learning is possible because of cooperation and knowledge building in cooperation. These information sources themselves can exhibit some sort of intelligence.

One of the most important issues faced in cooperative learning is the evaluation of the information and the knowledge acquired from other sources and incorporating that knowledge to build a system view.

The independent learning takes place in each intelligent agent. These agents interact among themselves for cooperative learning.

Figure 4.5 depicts an SL model with reference to system boundary. SL takes input about the boundary and the system environment, and every action is tested with reference to a learning policy. The learning policy is refined with the response from the environment for that action.

The multiple validation and reassessment are done before the data are used for learning.

Systemic Machine Learning Models. SL models are of two types:

1. Interaction-centric models
2. Outcome-centric models

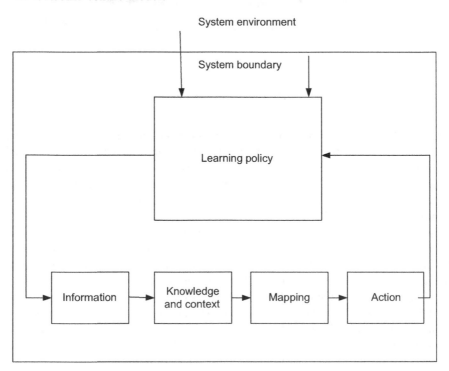

Figure 4.5   Systemic learning model.

4.2.2   Interaction-Centric Models

In interaction-centric models, the learning takes place based on the results of small interactions and not merely based on resultant outcome. The outcome-centric models derive learning based on outcome. These may look a bit similar to temporal difference methods in reinforcement learning. The major difference lies in the measurement of interactions and outcome. Interaction-centric learning is also referred to as cooperative learning. Cooperative learning refers to learning together. This allows the learning based on observation and information gathered by multiple agents. In cooperative multiagent learning, several agents maximize learning through their interactions. The systemic machine learning is inherently a multiagent system. Since SL expects understanding of interactions and behaviors among subsystems and different parts of a system, it is a sort of cooperative multiagent learning. Figure 4.6 depicts the interaction-centric model. Here a number of subsystems interact with one another, and these interactions are used for systemic machine learning.

4.2.3   Outcome-Centric Models

Outcome-centric models are the models basically based on outcomes. These models are not based on intermediate interactions. The outcome at any transition stage is used

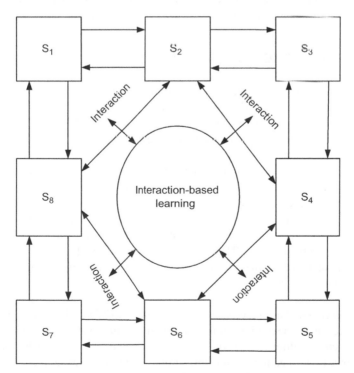

Figure 4.6   Interaction-centric model.

to infer the systemic parameters. Outcomes at various subsystems are used. In this particular case, the impact space is predefined. The relationships among subsystems are used to discount the impact on various subsystems in impact space. The learning is based more on outcomes and discounted impacts and inferred systemic decision parameters. It is more like reward-based learning used in the case of reinforcement learning. The reward is discounted. In this case, the reward is measured in impact space and discounted across the impact space to determine the context and the representative rewards.

The outcome-centric learning is simplified with outcomes. For actions or a set of actions, say $[a_1, a_2, \ldots, a_n]$, the outcomes for the subsystems $[s_1, s_2, \ldots, s_m]$ are, for example, $[a_{11}, a_{12}, \ldots, a_{mn}]$.

> *Representation of Outcomes*: For any action a_1 let s_1 be the outcome. These outcomes are mapped to action for learning.
>
> *Reward Calculation*: For every action outcome can be mapped with reference to decision scenario. The reward is calculated with reference to decision scenario. These rewards are used for learning.
>
> *Discounted Rewards*: These rewards cannot be uniform and need to be discounted with reference to time and relevance.

Figure 4.7 depicts an outcome-centric learning model. Here the learning results are not based on relationships among subsystems but are based on outcomes.

4.3 CAPTURING THE SYSTEMIC VIEW

The systemic view is the view of the system which represents the relationships among different parts of the system, subsystems, and their dependencies and gives the complete relationship and decision-centric picture of the system.

The system is generally fragmented. Determining the system boundaries and building a system view for decision making are two of the most critical tasks. The system view is captured in pieces and integrated to build a complete system view. The systemic view consists of horizontal and vertical views. The system boundaries are not known hence building system view is a difficult task. An intelligent agent or other data-collection sources collect information from different parts of the system and also from different perspectives. The system view is built based on the captured data. There are various ways that can be used to use this information to build the optimal system view. Three important models are discussed below:

- *Predetermined Boundary-Based System View Building*: In this case, data are captured within the predetermined or already defined system view boundary. For every action within the action space, the changes in data and behavior with reference to defined parameters are studied within these predefined boundaries. A typical predetermined boundary-based system view building is depicted in

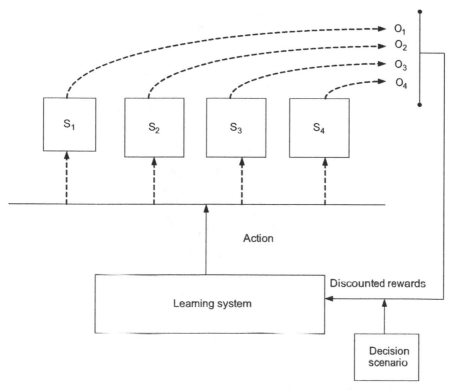

Figure 4.7 Outcome-centric model.

Figure 4.8. In this case, the boundaries of a system are predefined and hence learning is based on outcomes of subsystems within the predefined system boundaries.

- *Dynamic Boundary-Based System View Building*: In this model, dynamic determination of the boundary is done through the repetitive analysis. Data are captured at various levels.

There are two approaches that can be used in this case. One is a parameter-based approach where relevance among parameters is used to represent relationship. The second approach is that for every action the pattern of impact is traced to determine the effective system boundaries.

For example, in a company there are many divisions: account division, engineering division, R&D division, and production division. For SL, there is a need to understand relationships among these divisions. To understand these relations, we need to know rational system boundaries. A typical example of building systemic view with reference to rational system boundaries is depicted in Figure 4.9. Here S_1, S_2, S_3 are subsystems, while p_1 and p_2 are problem-specific perspectives for subsystems S_1; p_3 and p_4 are problem-specific perspectives for subsystem S_2. Similarly, p_5 and p_6 are

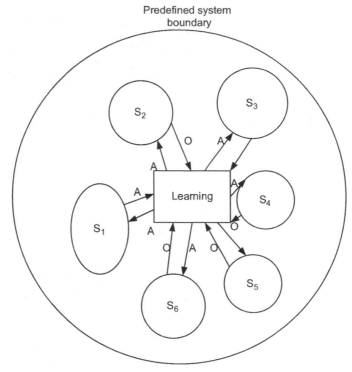

Figure 4.8   Predetermined boundary-based system view building.

problem-specific perspectives for subsystem S_3. The overall context is built as depicted in Figure 4.9.

This helps us in building a system-context view. "System-context view" refers to understanding of all parameters of the system in system context. Context view refers to the view that deals with connections and relationships among systems with reference to environment in a defined context. The concept view looks into internal boundaries with reference to visible decision outcomes and impacts. The context and concept view together build the complete system view. A typical concept view, context view, and the system view relationship is depicted in Figure 4.10.

The system view will define

Parameter set $\{p_1, p_2, \ldots, p_n\}$
The list of prime parameters: $(q_1, q_2, \ldots, q_m)$
Prime parameter set is a subset of parameter set.

Relationships among different parameters are defined in terms of closeness and impact. Parameters of interest and priorities are determined based on impact. Impact factor is used to define this impact.

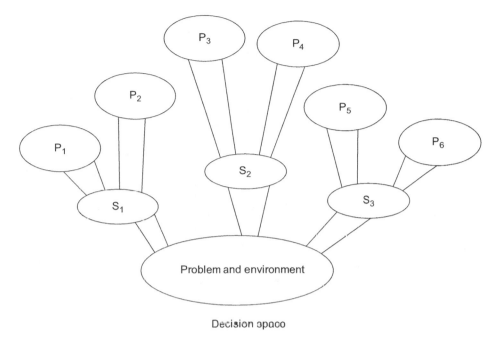

Figure 4.9   System view with perspectives.

4.4   MATHEMATICAL REPRESENTATION OF SYSTEM INTERACTIONS

The background knowledge with reference to interactions and historical data of inputs and outcomes needs to be used. The representation of knowledge about system interactions is very important from systemic perspective. Inductive learning is learning by examples. Knowledge-based inductive learning tries to infer the relationships based on prior knowledge. A system consists of different elements, different layers, and relationships among these layers and elements. These relationships are not known but can be inferred or determined based on prior knowledge.

Let us consider a system S. Based on prior knowledge, system S consists of n subsystems $S_1, S_2, \ldots, S_n$. Out of these subsystems, let S_A be the subsystem that is the action space for the particular decision-making scenario.

Now for some of the subsystems, the outcome parameters are available at $t = t_0$. For some subsystems, the outcome parameters are available at $t = t_1$, and so on. For some subsystems, no outcome parameters are available.

Dependency is initially represented in terms of direct connection, historically seen impact (knowledge), and closeness. The dependency is discounted beyond action

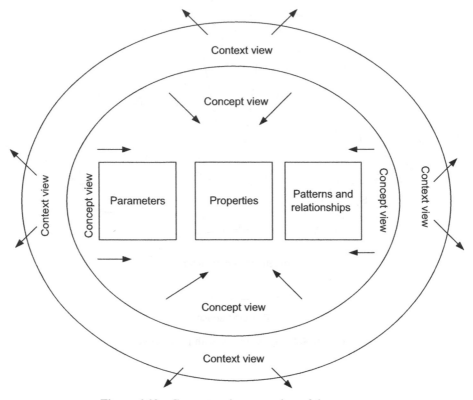

Figure 4.10   Concept and context view of the system.

space. The relevance of parameters is considered in the case of decision making along with dependencies. The system vector is built with reference to action space.

$$\text{Impact factor} = (p1/da1) \times \gamma \times \text{closeness}$$

Elements of decision matrix are calculated by

$$W_{ij}(\text{impact factor}) * P1 + \text{pattern weight}$$

The matrix given below represents a decision matrix. Here m is the number of parameters while n is the number of subsystems.

$$
\begin{matrix}
D_{11} & D_{12} & D_{13} & \cdots & D_{1m} \\
D_{21} & D_{22} & D_{23} & & D_{2m} \\
D_{31} & D_{32} & D_{33} & & D_{3m} \\
& & & & \vdots \\
D_{n1} & D_{n2} & D_{n3} & & D_{nm}
\end{matrix}
$$

While exploiting the already learned facts, SL uses past knowledge to determine impact factors. The past or learned impact factors are available for the actions that were performed in the past. The impact-factor determination through exploration takes place in case of new actions. Interestingly, neither exploration nor exploitation is sufficient to determine systemic relationships.

4.5   IMPACT FUNCTION

Impact function helps in calculating the impact of any decision on the environment. For any action the impact is seen not only within a system but in the environment and neighboring systems. Based on decision-impact analysis impact function is defined. The impact function is the approximation of impact of action and is based on different parameters and it helps in calculating the impact. The impact function helps us to determine the impact of any action on a particular parameter. Impact function is generally derived from the pattern of the impact.

4.6   DECISION-IMPACT ANALYSIS

The impact of any action based on a decision needs to be determined with reference to each of the subsystems. The action space is the space in which action is performed. The discount factor gamma equals 1 for parameters in action space. For each possible action in action space, the impact factor for all parameters is calculated. These parameters are represented as a matrix.

$$AS_1 = f(Ps_{11}, Ps_{12}, Ps_{13}, \ldots, Ps_{1n})$$
$$AS_2 = f(Ps_{21}, Ps_{22}, Ps_{23}, \ldots, Ps_{2n})$$

With reference to action A, all the parameters are prioritized with reference to impact. There are various ways to analyze decision impact. As defined above, the impact of any action over the set of parameters is observed, inferred, or calculated based on impact function. This impact analysis is continuous as the impact of the same action can be observed much later. The interesting part of this analysis is mapping the action and impact. Since in decision space there are many actions taking place, generally measurement of parameters continuously helps in mapping actions to impact. We need to know the impact matrix with reference to every action.

For example, in the time space under observation, action $\{a_1, a_2,$ and $a_3\}$ took place. We get a decision matrix for parameters within space and time boundary. Such series of actions and matrices help us to determine the relevance of any impact with reference to a particular action. That results in building the decision matrix for the action. Decision matrix helps to convert the data produced to the information required for systemic machine learning. There is an information gap between information needed and data produced or available. In some cases, data available may include all information required, but separating this information is quite a complex task. The decision-impact analysis helps us in separating the information required from the data available. Figure 4.11 depicts this typical scenario.

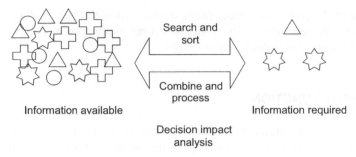

Information available

Search and sort

Combine and process

Information required

Decision impact analysis

Figure 4.11 Decision-impact analysis.

4.6.1 Time and Space Boundaries

As we have discussed in the previous section, time and space boundaries are considered while we are performing impact analysis for any action. These boundaries define the limit of relevance for learning. While learning, we consider all relevant parameters within time and space limit. The impact on these parameters of any action is not uniform, and that is determined through a system view and impact analysis. Figure 4.12 depicts the process of time and space boundary detection using impact analysis.

Figure 4.12 depicts time and space boundaries. Maximum likelihood can help in deciding the impact. The gradual increase in time and decision space helps us in

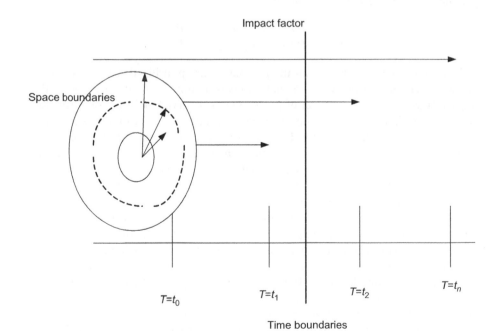

Impact factor

Space boundaries

$T=t_0$

$T=t_1$

$T=t_2$

$T=t_n$

Time boundaries

Figure 4.12 Time and space boundaries diagram.

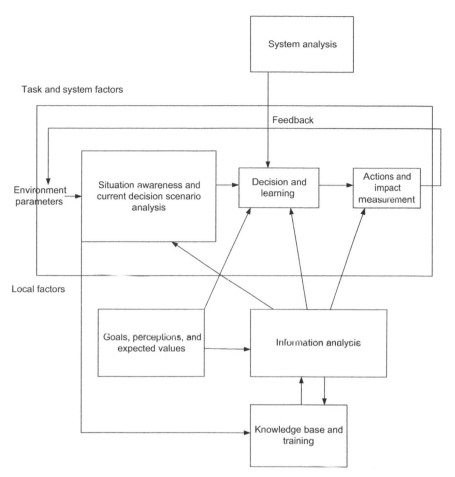

Figure 4.13 Systemic learning framework.

identifying the boundaries for decision scenarios. One of the special cases of SL is where space boundaries and decision boundaries are same. The time and space boundaries provide help to build a system view. This can be further used for decision making in various applications. Figure 4.13 depicts a complete framework for SL based on system boundaries.

Figure 4.14 depicts the relationship between environment, knowledge base, and SL. SL can be used in a number of applications. Figure 4.14 also depicts the role of systemic and cooperative learning for various applications.

4.6.1.1 *Example—Project Management and Systemic View* The system-ic view is very useful for learning and decision making. But it is more relevant in the case of applications that are dynamic, and there are a large number of parameters. Any action within the project or part of a component in the project might have an impact on other components—even on overall integration of components—and it may even

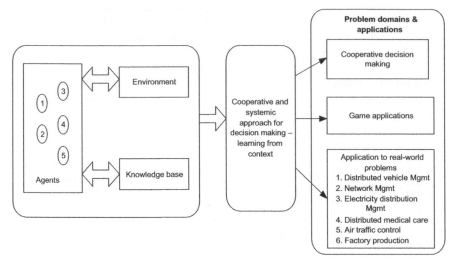

Figure 4.14   Systemic learning and applications.

indirectly impact on the date of final delivery of impact. So we can see that the impact of action or decision in a particular decision space might have impact beyond that decision space and visible time space. Identification of relations among different systems and subsystems within the projects, their dependencies, and relationships help in overall learning. Further understanding the possible considerable impact of action in time space might be until the end of the project or until we get the next assignment, which helps us continuously look at the decision and keep learning. The rewards in the case of any action are coming in the form of impact on various parameters in the system. This impact along with systemic view in time and space makes the systemic machine learning possible. For the project management, an SL framework is depicted in Figure 4.15. There are a number of information sources such as

Customers who want these products
Relevant industries
Previous experience
Feedback on similar products

Furthermore, there are a number of parameters such as

Quality
Milestones
Activities
Skills
Development environment
Work environment

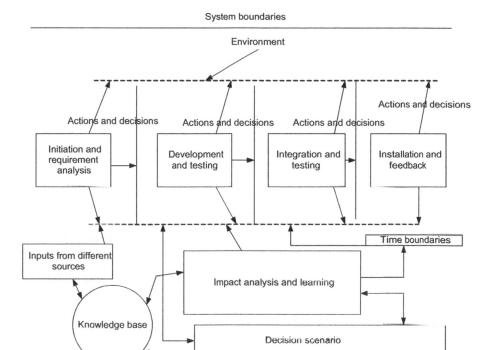

Figure 4.15 Systemic learning framework for project management.

For every action, impact analysis and learning with reference to decision scenario and knowledge base allows us to enrich the decision. The decisions are made with reference to system and time boundaries. These time and system boundaries are determined with reference to impact analysis of actions. Similarly, a systemic wholistic model for security informatics is depicted in Figure 4.16.

4.6.1.2 Systemic Model Development (Case Study) Systemic machine learning and decision making are useful in the case of complex systems and dependencies among different subsystems. A typical case study of a health-diagnostic system is given in this section. The diagnosis results from the information coming from different sources. There are many information sources such as

Lab reports
ECG reports
Lifestyle-related information
Hereditary tendencies
Specific habits
History of previous ailments

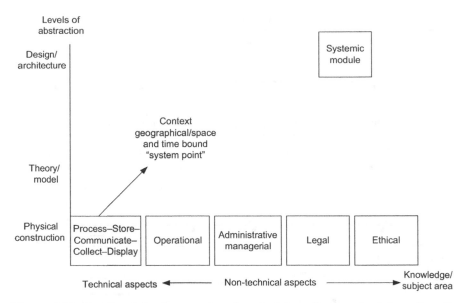

Figure 4.16 Details of the framework and methodology for security informatics—the systemic-holistic model.

Region of residence

Properties of geographical region of residence

Previous treatments and medications

Profession

There are many such sources of information that can help in diagnosis. The first step is information accumulation and prioritization. Each source of information leads to a vector. This vector represents the relevance and impact. These vectors together with environmental outcomes help in building a decision matrix. This decision matrix is used for learning and decision making. With every additional piece of input information, this matrix is modified to accommodate dynamic changes in system. One of the most important aspects is understanding dependencies among these parameters. Learning based on individual parameter fails to give the complete picture. For example, learning based on just blood pressure (BP) and dose of BP medicines can result in many side effects. The statistical analysis of all associated parameters and learning can cooperatively help to bring optimal solution and further to understand the impact of any decision. In this case, the system may be family or even close environment that will define space boundary while the time boundaries are expanded in future and past. System scope is depicted in Figure 4.17.

Figure 4.18 depicts the process of systemic decision matrix building for medical diagnostic system. The relevance of observed parameters with reference to decision problem is determined to confine system boundaries. This is performed by continuous dependency analysis and decision analysis. All these parameters are prioritized, and the relationship of these parameters is represented with reference to

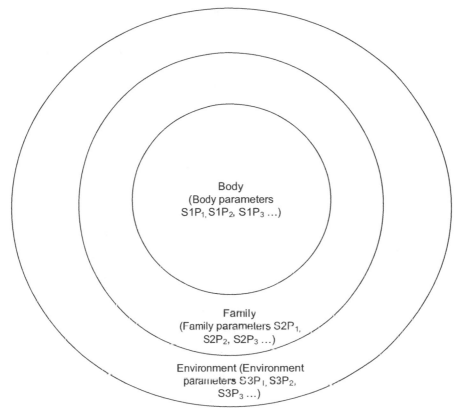

Figure 4.17   System scope.

decision problem. This leads to building a systemic decision matrix. This matrix leads to optimal decision.

4.7 SUMMARY

This chapter discusses one of the most important aspects of systemic machine learning, that is, model building. The systemic machine-learning model has a number of important parts including system boundaries, impact analysis, and building overall system view. In an SL model, the information is gathered from all available information sources. This information is imperfect information as it may be partial, may contain noise, and may also contain information that may not be relevant from the decision scenario. This information along with systemic inputs, patterns, feasibilities, impact, and possibilities are used to build knowledge. This knowledge and decision analysis helps to build the comprehensive decision scenario. Decision scenario and system information helps us for boundary detection. With the entire available information, the context for decision making is formed and used for systemic machine learning and decision making. The model has three important parts: knowledge base,

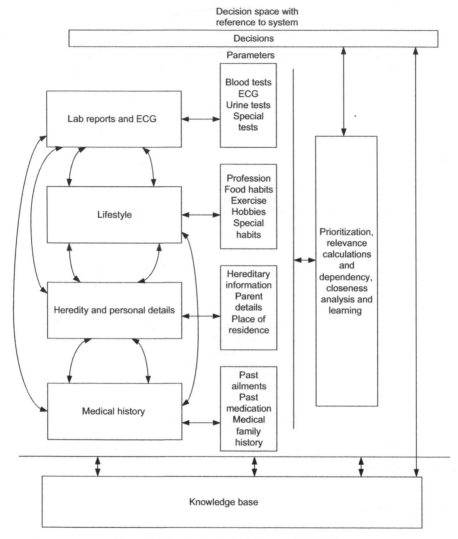

Figure 4.18 Systemic decision matrix building.

system information, and impact and relevance information. Learning can be represented in the form of a decision matrix. Since SL needs to be continuous here, the changes in outcomes and parameters at every time step are input for learning. The process is adaptive.

The relevance of parameters with reference to any action is determined based on impact analysis. The SL needs to understand the impact of any action and weight of different parameters in light of the decision scenario. All these concepts together build the holistic system view, which can give better insight for learning and can lead to better decision making.

CHAPTER 5

Inference and Information Integration

5.1 INTRODUCTION

Learning and decision making are based on the selection of the right information and effective use of available information. Learning is not that difficult when the complete and relevant information is available. But in most of the applications as we discussed in previous chapters, only limited information is available and the relevance of information is not known. In real-life problems, many times sample space is an abstract space. When the parameter space and sample space are abstract, it demands systematic inference for building the knowledge base for effective learning. This chapter deals with learning in the case of abstract parameter and sample space. The combination of pattern analysis and statistical inference is required to build the system view and relationship among available parameters. Most of the standard inference and data-building methodologies are based on probabilistic likelihood calculations. Maximum-likelihood estimation and Bayesian inference can be used for maximum-likelihood-based inference. Classical methods of inference can be extended to abstract spaces and even further for the abstract spaces from a systemic perspective.

In the case of real-life problems, partial and heterogeneous information pose various challenges when we try to combine and use this information. This limited information can be used effectively with inference to determine contextual aspects of decision. For any decision making, the complete information about decision scenario is necessary. All the information about decision scenario is rarely available. Even in many cases in real-life scenarios we have to deal with a lot of unknowns. Information coming from more than one source offers different perspectives, but unfortunately establishing relationships among these pieces of information is a difficult task. In a system there are many subsystems within system boundaries. All the parts of the system are not visible, and it becomes necessary to derive various parameters in the system in order to produce optimal results. The data-driven inference includes, simple methods such as interpolation, extrapolation, and many statistical-inference-based methods.

Reinforcement and Systemic Machine Learning for Decision Making, First Edition. Parag Kulkarni.
© 2012 by the Institute of Electrical and Electronics Engineers, Inc.
Published 2012 by John Wiley & Sons, Inc.

These methods are based on adjacent or close data points and perceived relationships among different data points. The need for inference becomes even more important in the case of bigger systems, unknown events, and abstract spaces. Generally, statistical-inference mechanisms are used in the case of all data-centric and pattern-based problems. Inference and available data can be used to determine the overall picture of a system and to generate a system view. In this process, there is a need for frequent data and information integration.

In this chapter, we will discuss various methods used for inference and information integration. The fusion of data coming from more than one source and cooperative inference are some other aspects of decision making we will try to cover in this chapter. Cooperative inference refers to inference based on data coming from different sources. Furthermore, cooperative inference can be used as a tool for mathematical idealization to approximate an observable phenomenon in a multivariable, complex scenario for systemic view building.

Various statistical methods, rule-based techniques, and pattern-based techniques used for inference are also discussed in detail in this chapter. Deterministic and nondeterministic models are used for inference. The inference mechanisms need to consider data, time of instance, dependencies, and context. Furthermore, those can be used to build context for the next level of inference. The continuous inference and use of the inferred information helps to build a better and better system view. As the underlying distribution is unknown it is nonparametric in nature. Time is another important aspect of systemic learning. As the cause and effect might be separated in time and space, there is need to infer possible data in longer future. This chapter will discuss inference techniques and use of inference mechanisms in system context.

Information integration that is also sometimes called information fusion is the merging of information from different and unidentical information sources with differing conceptual, contextual, and typographical representations. The form relevance and context of the data might be different. The major problem that arises in the fusion of data is diverse and uncertain information from different contexts which needs to be combined for a particular decision scenario. The parametric models used for data fusion in the case of dynamic scenario exhibit problems due to abrupt changes in parameters and unknown time instances. Since in all real systems data come from unstructured or semistructured resources, there is need for some sort of consolidation.

The technologies available to integrate information include string metrics, which allow the detection of similar text in different data sources by fuzzy matching. The statistical approaches such as Bayesian approaches and Markov Chain Monte Carlo (MCMC) methods can be used for information integration.

This chapter focuses on parametric and nonparametric inference techniques along with cooperative inference and data fusion. The overall systemic view building is like building the whole picture from the fragmented facts available in bits and pieces. It is like an FBI agent trying to build a complete case based on reports and investigations coming from different sources and witnesses with the use of previous experience, cooperative inference, and data fusion. At every moment there is new information coming, and new facts are revealed; this can give new dimensions for learning and decision making.

Statisticians distinguish between various modeling assumptions based on the level of complexity as:

- *Fully Parametric*: The probability distributions describing the data-generation process are assumed to be fully described by a family of probability distributions that has a finite number of unknown parameters.
- *Nonparametric*: The assumptions made in this case are that the number and nature of parameters are flexible.
- *Semiparametric*: This term typically implies assumptions "between" fully and nonparametric approaches.

Inference makes it possible to deal with the information gaps and build the overall picture for decision making and learning.

5.2 INFERENCE MECHANISMS AND NEED

Inference is the process of arriving at some conclusion based on available facts, which may or may not be logically derivable from the assumed premises, and is based on probability of certain events that are happening. Since there is a lot of uncertainty in real-life scenarios and a lot of new information is becoming available at every moment, it is absolutely necessary to come up with inference that can help in making the best choice. More relevant information and better mechanisms to build context help in making a better inference. The multilevel inference can help to slowly nail down the decision. Another most important part is cooperative inference based on multiple data sources and inference mechanisms.

There are a number of methods used for inference. Many of the most common methods are based on statistical inference. In a parametric statistical inference, data come from a type of probability distribution and are used to make an inference about parameters. Nonparametric statistical inference is one where it does not rely on data belonging to a particular distribution. In this case, the interpretation is independent of parameterized distribution. The structure of these models is not fixed but dynamic. Another very commonly used concept is transitive inference.

Transitive inference uses previous training relationships between stimuli to determine a relationship between stimuli that are coming or presented at different times of instances. Transitive inference is about producing appropriate responses to novel scenarios without the exact experience of that scenario. It is based on transition property and makes effective use of past training relationships.

Stimuli are events in the environment that influence behavior. A single stimulus can serve many different functions. Transitive inference, according to some [1–3], is based on deductive reasoning because there is a need to deduce or determine a relationship between stimuli that is not explicitly presented. From the single observation there are multiple inferences possible.

Inference does not merely follow the available data or pattern. It is rather made when a person (or machine) goes beyond available evidence to form a conclusion.

Human beings use this technique quite often. In cases such as investigation of crimes, detectives many times go beyond the available evidence. With a deductive inference, this conclusion always follows the stated premises. Even in the case of diagnosis of the ailment of a patient, the doctor goes beyond evidence available in the form of test reports. Deductive is a more logical inference mechanism and is very important for machine learning to learn from complex and incomplete inputs. Deductive as well as transitive inference are required to build the overall context. The inference is continuous and hence at every level of new parameter, information and data are inferred and are used in the next level of inference or iteration. Iterative inference is also used in many cases. Inference is generally indirect and hence depends on information that is coming from different sources. Hence one of the most important aspects is cooperative inference. Cooperative inference is based on cooperation or effective and logical use of information available from different sources.

One of the most common inference mechanisms is statistical inference. Statistical inference is the process of arriving at a decision or drawing conclusions from data and its variations. There are random variations in data. These data can be of system performance anomalies, system performance, or behavioral changes. The statistical inference and based procedures can be used to draw conclusions and derive decision from datasets arising from systems affected by random variation. Since a system with multiple parts has data available in various pockets and different formats and is also collected with different perspectives, there is need of data inference. One possible option that can be used for inference in such cases can be the statistical method. Initial requirements of such a system of procedures for inference and induction are that the system should produce reasonable answers when applied to well-defined situations and that it should be general enough to be applied across a range of situations.

Figure 5.1 depicts the role of the inference engine in a typical expert system with machine learning. Here the outcome is knowledge building and learning. The inference

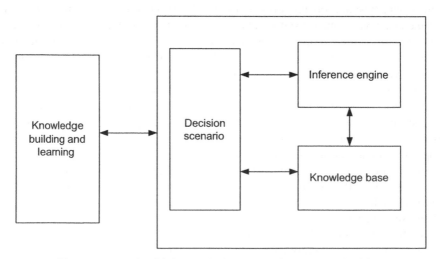

Figure 5.1   Role of inference in learning and knowledge building.

engine takes external inputs in the form of decision scenario and also interacts with a preexisting knowledge base. Furthermore, it builds and strengthens the knowledge base. The decision scenario and knowledge base helps in building knowledge and inferring new information to make logical learning possible.

5.2.1   Context Inference

Context is the overall scenario and environment in which the action, decision, or relationship is being executed. To determine the context for decision making, there is a need for context inference. The same decision scenario may lead to different decisions in different contexts. Context inference refers to the process of inferring systemic parameters to build context for decision making. Thus context inference is the process of inferring the overall decision contexts based on data coming from multiple input data sources, available knowledge base, and system information. Determining context hierarchy, sharing relevant information, and use of the right algorithm to infer overall systemic context are some of the challenges faced during context inference.

Figure 5.2 depicts the use of context inference for decision making. Data inference is more statistical, while context inference uses a cooperative inference mechanism where information from multiple sources is used. The decision data and available information are preprocessed. The feature and parameter sets along with mapping among them helps in prioritization of features. The features, parameters, mapping among them, and the prioritization can help in building an inference model and formation of rules. This makes the context-based learning possible.

5.2.2   Inference to Determine Impact

We have discussed in the previous section that for systemic machine learning, understanding the impact of action with reference to a system is very important. Furthermore, knowing the system boundaries and interrelationships helps in understanding this

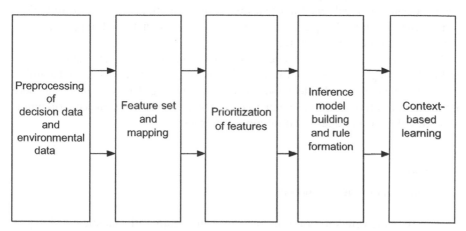

Figure 5.2   Inference approach for context-based learning.

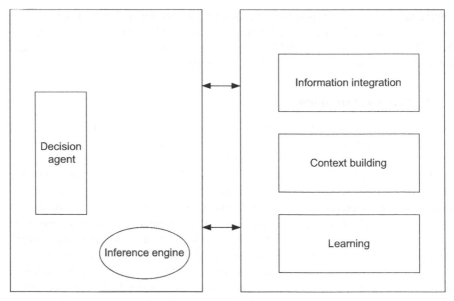

Figure 5.3   Learning architecture.

impact. One of the most important aspects of systemic learning is establishing relationships between cause and effect that are separated in time and space. Traditional machine learning is dependent on immediate and relevant data, which is directly deducible with reference to any action. That does not mean that systemic machine learning should ignore the direct and immediate impact of action. For any action there are two types of impacts that can be observed: (a) direct and visible or rather directly deducible impacts and (b) indirect and inferred impacts.

Inference can be used to determine both impacts and can be given weights to select the right or the most appropriate actions. In the subsequent section, we will discuss in greater detail statistical and Bayesian inference and how it can be used to build a system view. Figure 5.3 depicts the framework for determination of impact based on cooperative and context inference.

As shown in Figure 5.3, inference based on actions, outcome, and data is used to build a knowledge base. The information integration, context building, and learning are in communication with a decision agent. This knowledge base with more and more inputs and transitive as well as deductive inference for any decision scenario builds context. Other knowledge-based inference includes

- Bridging
- Contrast
- Understanding new expressions

Here bridging refers to bridging the information gaps with statistical techniques or with the help of information available from other sources. The contrast and

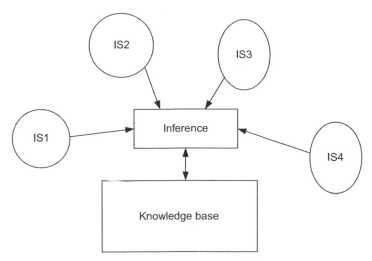

Figure 5.4   Knowledge–inference relationship.

understanding of new expressions helps in accommodation of new information to build relevant information. Knowledge-based cooperative inference takes knowledge built from various sources. Logical and relational learning builds the foundation for cooperative machine learning. Figure 5.4 depicts the relationship between inference and knowledge base. The information coming from multiple information sources IS1, IS2, IS3, and IS4 is used to infer the facts with inputs from knowledge base. The inference provides input to build knowledge base. As we discussed previously in the case of FBI, IS1 is the information coming from doctors and investigations, while IS2 is the information coming from prime witnesses, IS3 is the information coming from the other witnesses, IS4 is the information that became available from crime scene, and so on. The information coming from all sources can be used during the cooperative inference. While doing this the knowledge built in the past and pre-existing knowledge are used to build the complete crime picture that helps in learning and decision making.

Figure 5.5 depicts the context inference framework. Inputs from different sources come in the form of raw data. These data are in heterogeneous forms and are also incomplete. Statistical inference can be used to take these data to the next level. Context inference takes inputs from the decision scenario and these data to build overall context.

Context inference and context access control are two important aspects of context determination. Context-source management tries to collect context-related information from the most relevant sources. Initially these sources are based on direct inference; but as more and more information becomes available, advanced inference rules are used to determine the context sources. Context modeling represents entire context data. Typically, context data are of two types: static context data and dynamic context data. Static context data are determined using simple rule-based inference, while dynamic context data are determined using dynamic inference. For distributed context management we need distributed inference based on limited information.

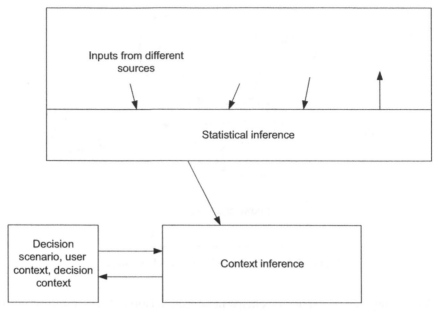

Figure 5.5   Context inference.

Dynamic inference is time based and hence considers time as a parameter for management. Context source management includes discovering context sources. Inference also considers preference analysis and history-based context inference and mapping among different parameters. The system and subsystem behavioral patterns are also used for inference. Context inference is the extraction of systemic or high-level context from the raw local contextual and parametric information. For context inference there is a need for dynamic inference and learning of context inference rules (CIR). These inference mechanisms build a platform for context association.

Cooperative inference and cooperative context inference are based on information coming from various sources, systems, and subsystems. The relationships among them allow the cooperative inference to build overall system level or high-level context inference.

Figure 5.6 depicts a typical cooperative inference. Context source, information from various sources, and mechanism on working cooperatively on different information sources allow building overall context. There are multiple levels of inference possible. Generally any level of inference allows taking the lower level available inference to the next level based on new available information. Cooperative inference goes through multiple iterations. The new inference changes the information and relevance of the already available information, and this further helps in broader context.

High-level context inference is system centric. This results from different levels of inference mechanisms. Integration of context and inference builds a system view. In the next section, we will discuss the integration of context and inference.

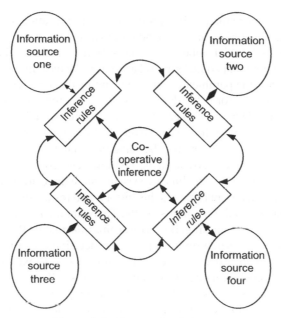

Figure 5.6   Cooperative inference.

5.3 INTEGRATION OF CONTEXT AND INFERENCE

Context and inference are integrated to derive the best possible decision. The integrated system does not necessarily exhibit the same behavior as prescribed by the inferred models of each component in the system. The integrated model should determine system boundaries and the possible impact of any action. Multiple sensors and agents need to be used to infer context. Here we will refer to them as information sources. Lower-level direct inference is transformed into higher-level systemic inference, which is referred to as context inference. The techniques used include

- User and decision scenario driven inference
- Probabilistic and statistical inference
- Rule-based inference
- Temporal logic inference

The overall decision matrix and decision making may change based on inference with adding of a new context dimension. With every new information, data, or change in inferred facts, the overall context may change. In a dynamic environment the parameters are changing. New context can be inferred based on the existing context in light of new information or new inferred facts. In some cases, static rules can be used. If a number of higher-level contexts are limited, rule-based methods can be preferred because their complexity is on the lower side.

Context refers to more than visible properties such as location. Context is all that may be used to characterize the surrounding environment, such as in the case of employees—it could be their hobbies, strengths, friends, residential address, and so on. Actually everything that could help to characterize the user directly or indirectly is the context. The data, the parameters, and the mapping that characterize the decision scenario along with the actor in the decision scenario together form decision context. Many decisions and decision makers can benefit from overall context that is the information beyond visible information. Every additional parameter and new available authentic information may help in building the context. To use this information coming in the system effectively, we need inference and context inference.

Let us take an example of a health care system where there are a number of inputs including a patient's health parameters, his history, and results of various tests. Now he has a particular complaint about his health. There are many parameters available while a few parameters are missing. Based on all this information the inference is used to build the complete information for a particular decision scenario. This complete information, along with a decision scenario, builds the context for learning. The process of building context using all the information is called as context inference.

Context inference, as discussed above, is building high-level context information from raw context data. This context building is based on mapping or based on learning of CIR that are based on algorithms for context building. It uses context association and pattern extraction/matching, along with CIR learning from cooperative and group knowledge.

The context management with reference to decision scenario—that is, priority and weight analysis, evaluation of priority, and continuous evaluation of priority with reference to environment condition monitoring—helps in building and establishing context. Context inference include knowledge-based context inference and history-based context inference. The context propagation is used for determining and handling complex scenarios. Group or cooperative context determination can be used to build a system context with reference to the available multiple contexts.

Figure 5.7 depicts the context building where a decision scenario is used by a learner to build the context, and this context is used for decision making.

Figure 5.8 depicts various levels of inference mechanisms. They are

- Data inference
- Context inference
- Decision inference
- System inference

Data inference is used directly on the extracted inference. The context inference is more cooperative in nature and uses data from different sources and builds context based on that. Decision inference uses decision-specific information and parameters relevant to decision scenario. The inference is made with reference to a decision scenario. The decision context is used for decision inference. The system inference is the inference made with reference to systemic context. The decision inference

Decision scenario and
environment

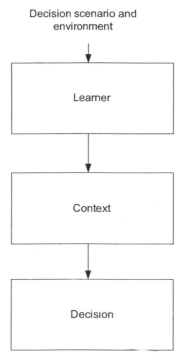

Figure 5.7 Context building.

and context inference is used for system inference. Simple statistical inference algorithms are used for data inference. Context inference uses cooperative inference algorithms. Decision inference uses decision inference algorithms and we need to use systemic inference algorithms with cooperative inference and systemic parameters for system inference. System inference is a sort of integrated context. Context fusion and context inference are used for building an integrated context.

Let D be a set of decision parameters for a system. Decision parameters are the parameters that play a role in decision making.

$$D = \{d_1, d_2, d_3, \ldots, d_n\}$$

For any two data parameters, a similarity index can be calculated using simple likelihood techniques. This similarity index helps in integrating the information. Let W_i be the weight of the ith attribute. The information extraction and compilation include information coming from different sources.

$S_1, S_2, S_3, \ldots$, and S_n are the sources of information.

Different sources of information give a subset of the parameter set. Decision parameters are a subset of attributes that are relevant for decision making.

$$D \in A$$

Here A is the set of all attributes. All attributes may not be known.

The decision attributes are prioritized with reference to a decision scenario.

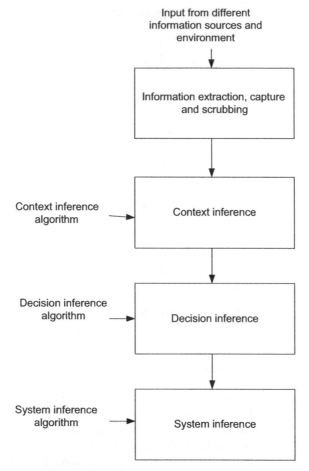

Figure 5.8 Context and system inference.

Source S_1 gives a set of values of decision attributes SD_1 while source S_2 gives a set of values of decision attributes SD_2 and so on.

The similarity score between required "decision attributes" and corresponding available attributes is calculated. These decision attributes are available from different information sources. The selected set of decision attributes, which is a subset of available attributes, is used to infer the complete set of attributes.

$$SE_i = \frac{\sum_{i=0}^{n}(W_i \times \text{Closeness}\,(SD_i, DA))}{\sum_{i=0}^{n} W_i}$$

Here W_i is the weight of the ith attribute, and closeness represents the similarity and closeness among the available attributes (SD) and required attributes (DA).

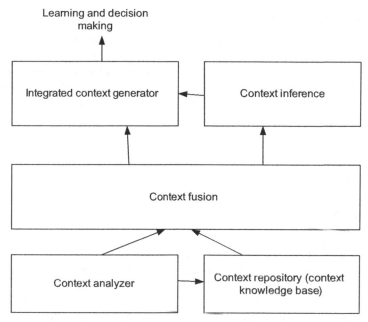

Figure 5.9 Context integration.

This closeness helps to select decision attributes for decision making. This leads to a set of values of decision and learning attributes. These attributes are used to build context. Figure 5.9 depicts the process of context integration and its use for learning.

5.4 STATISTICAL INFERENCE AND INDUCTION

Statistical inference is inference or drawing conclusions based on data. In different context inference is desirable. There are parametric, semiparametric, and nonparametric statistical inference models. In parametric models, a data-generation process is assumed to be described by finite unknown parameters. In a nonparametric one, the number and nature of parameters are flexible.

One of the most commonly used statistical-inference mechanisms is the likelihood technique. Another commonly used inference technique is point estimation. Point estimation makes a reasonable guess for missing values. In some cases where direct inference is useful, this method works very well.

5.4.1 Direct Inference

Direct inference provides a link between personal judgments and the available information about objective probability. These are generally based on calculable mathematical probability.

5.4.2 Indirect Inference

Indirect inference is an experiment-based, simulation-based, or outcome-based method for estimating, or making inferences about, the parameters. It is most useful in estimating models for which the likelihood function is known. These parameters can themselves be estimated using either the observed data or the simulated data. Indirect inference chooses the parameters of the underlying model.

5.4.3 Informative Inference

A body of data beyond mathematical likelihood based on information is referred to as informative inference.

5.4.4 Induction

Induction refers to guessing or logically determining the true underlying state of the world from limited observations and from incomplete information coming from heterogeneous sources. Bayesian inference and similar inference techniques can be used to infer about the underlying structures and relationships based on a set of limited information and available partial picture. As per Karl Pearson (1920), induction refers to the estimation of results in subsequent samples by using uniformity and representativeness assumptions [4].

5.5 PURE LIKELIHOOD APPROACH

The pure likelihood approach gives statistical-inference procedures. Here the collection of evidential outcomes that can be represented as statements is based wholly on the likelihood function. Likelihood represents the probability of high occurrence or the natural occurrence of preferences among the probabilities under consideration. This means that the approach satisfies the strong likelihood principle. In a sense the likelihood function is at the core of determination and gives evidential meaning of the statistical evidence. Even uncertainties in evidential statements are represented using conditional principles.

The value of the outcome, say X, is a dependant of some unknown parameter, say θ. This depends on the underlying model. Parameter estimation is done by defining a model and constraints based on domain knowledge and then solving for most likely values for the parameters of the model. We are given the observed data and the model of the interest. Then find the most suitable probability density function that will most likely produce the observed data. For this purpose the likelihood function is defined as

$$L(w|y) = f(y|w)$$

Here $L(w|y)$ represents the likelihood of the parameter w given y, where y is the observed data. Once we have the data and likelihood function, statistical inference about the population is possible. Maximum likelihood estimation was developed by Fisher [5]. Desired probability function is one, which makes the observed data most likely. This in short means this probability function must get the value of a parameter vector that maximizes likelihood function $L(w|y)$.

The values of a number of parameters are observed. The likelihood of observing given data with reference to selected decision parameters is calculated. Let us assume that parameter is p then $\text{lik}(p)$ = probability of observing given data as a function of p.

$$f(x_1, x_2, \ldots, x_n|p)$$

Hence the likelihood can be represented by

$$\text{lik}(p) = \prod_{i=1}^{n} f(x_i|p)$$

The maximum likelihood of p is the value of p that makes the observed data most probable. There is the observed behavior of the system and parameters as opposed to expected behavior. In a system, for any action, the natural order of the possible impact can help in determining the closeness among the systems and subsystems.

The likelihood approach is about the use of a collection of evidences and evidential statements and further uncertainties of these evidential parameters are calculated.

5.6   BAYESIAN PARADIGM AND INFERENCE

Bayesian paradigm of inference is based on conditional probability, and it is based on Bayes' theorem.

5.6.1   Bayes' Theorem

The best model or the preferred model is the model that maximizes the probability of seeing the data similar to the data that is observed. The Bayesian approach of inference represents this conditional probability $\Pr(Z|\theta)$. Furthermore, the Bayes' theorem helps us represent updated knowledge about θ after the observed data are given.

If the prior distribution is given, the posterior distribution can be determined using Bayesian theorem. Let pr be the prior distribution while po is the posterior distribution.

$$p(po|pr) = \frac{p(pr|po)p(po)}{p(pr)}$$

where

$$p(pr) = \int p(pr|po)p(po)\, dpo$$

and

$$p(pr|po) = \prod p(x_i|po)$$

The posterior distribution can be used to estimate po.

The inference techniques introduced above are the standard techniques to determine the maximum likelihood. The repetitive use of the standard likelihood techniques allows us to decide the weights of parameters in a system. Furthermore, this helps us to determine the system boundaries.

So $P(I|A)$ represents the probability of substantial impact on a set of parameters given that the action A is performed.

$$P(I|A) = \frac{P(A|I)*P(I)}{P(A)}$$

That is further extended in a simple way to represent the posterior distribution to understand the impact

$$P(I|A) = \frac{P(A|I)*P(I)}{\int P(A|I)*P(I)dI}$$

This posterior distribution also provides a platform for the prediction of impact for future actions.

5.7 TIME-BASED INFERENCE

One of the major aspects of systemic machine learning is that cause and effect can be separated in time and space. For any action there is an impact. This impact might be spread over the time period. Understanding this impact on relevant parameters in the system within a relevant time frame is required for learning. Time-based inference aims to determine the impact of the action on the scale of time.

Furthermore, there are a series of outcomes, and these outcomes impact future outcomes. As we discussed in the previous section, Bayesian inference gives us posterior distribution and also provides a platform for predicting values for future observations. Now we are interested in understanding the relationship among impact and actions at various time instances. So let us assume that I_{t_n} is a future observation at time instance t_n. We are interested in determining $P(I_{t_n}|A_{t_0})$. For a given action A at time $t = t_0$, what is the probability of impact I at time instance t_n?

$$P(I_{t_n}|A_{t_0}) = \int P(I_{t_n}|I_{t_0})P(I_{t_0}|A_{t_0})\, dI$$

The most difficult part is understanding the cumulative systemic impact of a given action.

5.8 INFERENCE TO BUILD A SYSTEM VIEW

To build a system view, various inference outcomes need to be combined. There are various parameters and aspects of the system which are inferred based on available data at various levels. Typical parameters and parts of inference include

- The boundary of a system
- Various subsystems and relationships among them

- Different possible action points
- Various possible actions
- Impact of possible action on various key parameters
- Action relationships
- Time-based inference to determine outcome

With all these inferences, an inference matrix is formed. This inference matrix helps in building the system view.

5.8.1   Information Integration

Information integration is the method of compiling and merging of information from disparate sources with different conceptual, contextual platform, and representations. As in systemic learning, we use data from different and disparate sources, and the need of consolidation of data from unstructured or semistructured resources arises. Information integration helps in representations of knowledge. Information and data from different sources with a few localized inferred facts are brought together and integrated in context with a decision scenario. The information available and inferred from different sources is integrated with reference to the context of the decision scenario. Let us take an example about learning for improvement of an educational system—here there are many parameters such as

- Properties of educational system $(p_1, p_2, \ldots, p_n)$
- Subject offer $(s_1, s_2, \ldots, s_m)$
- Methodologies followed (M)
- Hours spent (H)
- Exam system (ES)
- Success and occupation of past students (SPS)
- Research methods used (RM)
- Application of learnt facts (LF)

Now the information comes from various sources such as

- Schools
- Education department
- People
- Past students
- Industries
- International colleges
- News channels

This information might be in different formats and needs to be integrated. The decision about the selection of educational institutes based on system ranking can be

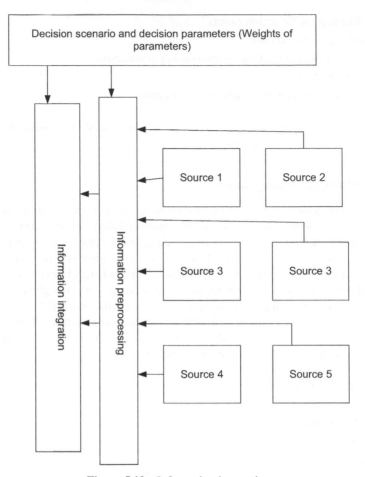

Figure 5.10   Information integration.

based on all these parameters with reference to a decision scenario. For example, if we are interested in a research-related degree, the weight of parameters may be different as compared to a business-related degree. Figure 5.10 depicts an information-integration process. The information coming from the multiple sources is preprocessed and then integrated with reference to a decision scenario and decision parameters to build integrated information.

5.8.1.1 *Selective Knowledge Building During Learning*   All the data and inferred facts may not be relevant in a given decision scenario. Knowledge building is not generalized, but rather it needs to be selective. The selective knowledge building is a collaborative one. The distributed information sources bring fragmented information, and we need information from a particular perspective. The learning process tries to explore a particular aspect of decision making and for this purpose tries to build

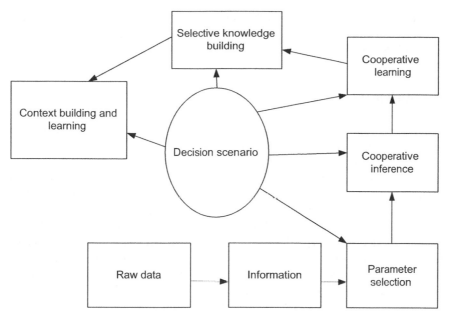

Figure 5.11 Selective knowledge building.

selective knowledge. Figure 5.11 depicts selective knowledge building with reference to decision applications. The information integration deals with compiling information from disparate sources and coming in different forms. The integrated information and inference helps in selective knowledge building for decision making. Raw data are converted to information. This information is used for parameter selection. A decision scenario plays a role in prioritization of decision parameters. Parameters along with the decision scenario are used for cooperative learning and cooperative inference. The decision scenario along with its context is used for selective knowledge building.

The context helps in building the selective knowledge for decision making. Figure 5.12 depicts the process of high-level context building. There are many

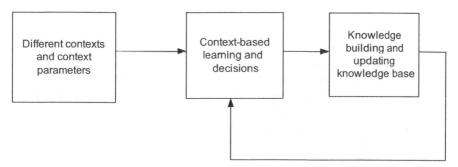

Figure 5.12 High-level context building.

contexts available. Context classification is used to determine the relevance of context. Knowledge is built continuously, and with reference to different contexts the knowledge base is updated.

5.9 SUMMARY

This chapter deals with one of the most important aspects of machine learning and systemic machine learning, that is, inference. When complete information is not available and when we need to make a decision in some scenario where exact facts are unknown, we need to infer data and information and build knowledge. Parametric, nonparametric, and semiparametric inference are the well-known three types based on complexity and use of distribution. We need to build a context for systemic decision making. In systemic machine learning, learning closely follows a decision scenario. The decision in a particular decision scenario demands different parameters, while a similar decision in another decision scenario may not demand these parameters. Inference is about data, and it may need to handle time and space constraints. In systemic learning, inference is finally used to build a complete system view with reference to scattered and fragmented information. Information integration and selective knowledge building can be used in a due process. Statistical inference techniques such as likelihood models, including Bayesian inference, can be used to determine the likelihood of impact of action in time space and decision space. This helps in deciding the system and time boundaries of interest.

The inference helps in building the overall context. This context helps various information relationships along with historical data. The inferred behavior of the system and response for actions with reference to parameters of interest allows us to determine the system boundaries. Irrespective of the technique and algorithm, the focus of systemic machine learning remains understanding the system, dependencies, and boundaries. The integration of data with reference to a decision scenario allows selective knowledge building for decision making. The learning for a particular decision scenario with a complete system in view makes systemic learning possible.

REFERENCES

1. Dusek J and Eichenbaum H. The hippocampus and memory for orderly stimulus relations. *Proceedings of National Academy of Science, USA,* 1997, **94**(13), 7109–7114.
2. Lazareva O and Wasserman E. Effect of stimulus orderability and reinforcement history on transitive responding in pigeons. *Behavioral Processes,* 2006, **72**(2), 161–172.
3. Ayalon M and Even R. Deductive reasoning: In the eye of the beholder. *Educational Studies in Mathematics,* 2008, **69**, 235–247.
4. Pearson K. The fundamental problem of practical statistics. *Biometrica,* 1920, **13**, 1–16.
5. Fisher R. A mathematical examination of the methods of determining the accuracy of an observation by the mean error and by the mean square error. *Monthly Notices of the Royal Astronomical Society,* 1920, **80**, 758–770.

CHAPTER 6

Adaptive Learning

6.1 INTRODUCTION

Adaptive machine learning refers to learning with skill of adaptation with regard to the environment, a decision scenario, or a learning problem. The learning is based on the gathered information, past knowledge, experience, and expert advice. A particular method that is very effective in a particular scenario may not be that effective in all the types of learning as well as in the decision scenarios. Human beings use different methods and learning strategies for different subjects, different situations, and for different problems. The method that is used for learning mathematics may be completely different from the method that is used for learning languages. Similarly, the method used for learning science may not be that effective if used while learning history. Furthermore, methods used for learning transforms and image processing may not work that well for geometry. The learning process is closely associated with the learning problem or rather what we are trying to learn and the learning objectives. Hence, selection of learning methods demands an understanding of the learning problem. There is a need to analyze the learning problem and then select the right approach or rather the most suitable approach dynamically in adaptive learning. It is not just switching among different methods or combining more than one learning methods. It is about the intelligent selection of data and the selection of the most appropriate method. It also involves dynamically changing the parameters or adapting to information to make the best use of data, namely, information with reference to a decision scenario. So, dynamic and adaptive learning is about amending learning methods and policies with reference to learning scenarios. It depends on actual user environment and the scenario presented. For two different scenarios, it also needs to consider relationships among those scenarios and different entities and parameters in those scenarios.

6.2 ADAPTIVE LEARNING AND ADAPTIVE SYSTEMS

An *adaptive system* is a set of various entities, which are independent or interdependent, real or abstract, forming an integrated whole system together that is able to

Reinforcement and Systemic Machine Learning for Decision Making, First Edition. Parag Kulkarni.
© 2012 by the Institute of Electrical and Electronics Engineers, Inc.
Published 2012 by John Wiley & Sons, Inc.

respond to environmental changes or changes in the interacting parts. Here the learning environment, data, and decision problem are changing. A learning system that can respond to changes in the environment and the learning framework to learn effectively is an *adaptive learning system*. In an adaptive learning system, the learning approach, weights of parameters, and selection of knowledge base are adapted with reference to the specific learning scenario. In short, the overall learning is based on the decision scenario and the information available. The course of learning and approach is more dynamic and adapts to the changing scenario.

Ensemble learning is a paradigm in machine learning where more than one learner is trained for the same problem. In the traditional approach, only one learner and a predefined approach with a single learning hypothesis is used for learning. The intelligence is sometimes used for determining decision-making thresholds. In the case of ensemble learning, a set of hypotheses is established. Many different learning approaches can be combined in the case of ensemble learning. The idea of adaptive learning is also based on the number of hypotheses. The concept of ensemble classifier is based on a set of classifiers that work on different hypotheses. Multiple experts, boosting, and voting are some of the techniques to use multiple learners for ensemble learning.

While learning, human beings do not stick to a single learning hypothesis, strategy, or methodology. Even for a single task, he/she uses multiple hypotheses such as ensemble learning and, most importantly, as a response to the environment switches among different strategies dynamically. There can be a number of expert opinions, different inferences, and acquired knowledge based on experience—all these things can be used effectively, appropriately, and based on needs. A typical multiple experts' scenario and formation of feature vectors used to come up with Meta expert or decision making is depicted in Figure 6.1.

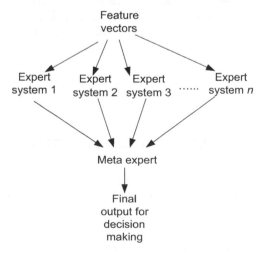

Figure 6.1   Meta expert scenarios.

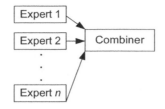

Experts operating in parallel

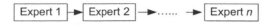

Experts operating serially

Hierarchical model of experts

Figure 6.2   Different multiexpert scenarios.

These experts can be in series or can be parallel. Each expert has relevance and can be represented with a weight associated with that expert. In the case of multiple expert scenarios, various approaches are used for making effective use of the knowledge from multiple experts. For technolegal problems we may need technical as well as legal experts, and weight of opinion from them cannot be simply combined but it needs to be tested in light of the decision scenario.

Various possibilities of multiexpert scenario for adaptive learning are depicted in Figure 6.2. These possibilities are based on the decision scenario and the need of multiple experts. The adaptive learning is not merely having more than one expert and a method for combining their decisions. It is actually adapting the learning policy based on the problem. Multiple learners and their combination is definitely one possibility to make adapting possible. A decision scenario may need the separation of a particular type of class or it may need relationship among different classes.

A typical complex decision scenario with reference to classification is depicted in Figure 6.3.

Adaptive learning also uses multiple learners with the ability of dynamic selection, learning based on response from environment, and dynamic analysis of scenario. There can be different learn sets, a set of algorithms, and these can sit on top of basic learners and classifiers. This combination along with the decision scenario will form the adaptive learning algorithm. The trickiest part in this type of framework is the

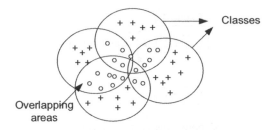

Figure 6.3 Complex decision scenarios.

ability to combine different classifiers and learners. A typical framework using multiple learners and learning algorithms is depicted in Figure 6.4.

In this chapter, we will discuss in detail ensemble learning, adaptive learning, and the role it can play in systemic machine learning. Adaptive learning needs to consider time period. Adaptive learning and adaptation of and algorithm based on a time line is depicted in Figure 6.5. The adaptation and selection of learning strategy is based on decision and learning scenario. The state of the system may change over time, resulting in a new decision and a new learning scenario. This state transition may take place due to changes in environment or due to the availability of new data.

The framework for adaptive learning should be able to establish mapping between the decision scenario and the algorithm, learning policy, and decision matrix. An adaptive learning model attempts to describe dynamic behavior and the mapping of

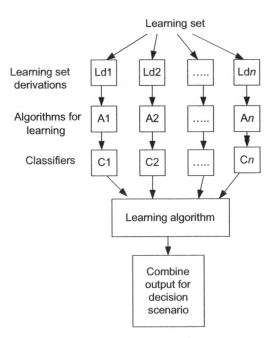

Figure 6.4 Multiple learners' framework.

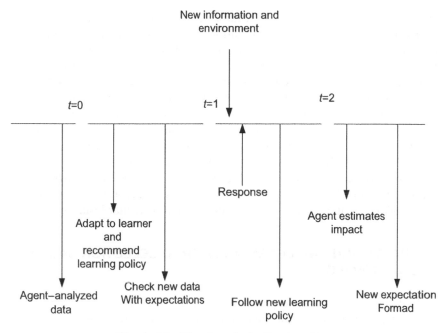

Figure 6.5 Time-based adaptive learning.

behavior and learning policy of an agent with changing decision problem and scenario. These models can be used in different learning scenarios and disparate environments. Hence the most critical part is the understanding and representation of the decision scenario and further mapping or deriving of the corresponding learning scenario.

6.3 WHAT IS ADAPTIVE MACHINE LEARNING?

The typical learning environment is dynamic. In real life, decision scenarios and the learning environment is changing continuously. There are changes in

- Environment
- Decision scenarios
- Perspectives
- Parameters

This results due to changes in parameters, availability of new information, and some external factors. We come across new information that has the potential to build a new knowledge. This knowledge may provide a new dimension for learning.

These changes pose the need of a dynamic learning method that can adapt to the environment to exhibit the best learning policy. In case of adaptive learning systems, the learner is no longer a passive receptor of the information but is in search of information, collaborates among information, and adapts to learning environment. It is a holistic framework for improving learning and decision making.

Since there is a changing environment and even the scenario and objective of certain decision making is context specific, the adaptive learning continuously interacts with the environment and adapts to the best possible learning algorithm and policy. One of the simplest forms of adaptive learning is the use of a number of learning algorithms or classifiers. There are many learning algorithms, and there can be a number of learning policies. No learning algorithm or policy can be effective in all possible scenarios. Hence it is a challenge to come up with a policy that can always produce the most accurate results and cope up with the dynamic environment. Even in the case of a single domain based on data and decision scenario, different learning policies and different algorithms are appropriate at a particular instance.

6.4   ADAPTATION AND LEARNING METHOD SELECTION BASED ON SCENARIO

The ability to reuse learning resources from large repositories, taking into account the context and to allow dynamic adaptation to different learners based on decision scenarios and learning policies, is carried out in adaptive learning. In adaptive learning, integration of different scientific approaches takes place. There are various adaptation models. Decision for resource selection according to the activity and context or rather some dimension is allowed in a few models. The learning domain and the knowledge domain may be kept constant in few models. Based on the degree of adaptation, learning models can be divided into three types:

Nonadaptable models
Partially adaptable models
Fully adaptable models

Adaptive learning enhances learning and performance of a learning system by using the fly inference, selectively by using distributed knowledge sources, and having a clear understanding of current context and relevance as well as the need of solution. The context awareness and adaptation of learning allows learners to deal with dynamic scenarios. Figure 6.6 explains learning with multiple learners using sets of parameters p1 and p2.

Furthermore, adaptive models are classified as interactive adaptive models, parametric adaptive models, adaptive ensemble models, and real-time adaptive models. In parametric adaptive models, learning strategy is updated based on observed parameters. Number of parameters, parameter clusters, and learning behavior are tracked in this model. In an interactive adaptive model, the parameters used for learning are determined based on continuous interaction with the system. The model interacts with the system to determine the learning strategy. In an ensemble model, several models are combined. Typically a mixture of experts, bagging, and multiagent systems can come under this category. In an adaptive ensemble model the ensemble model is selected

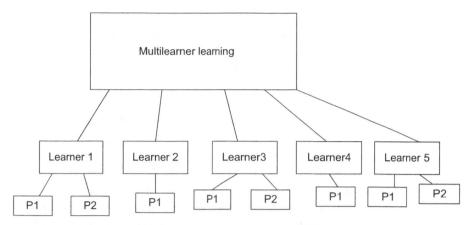

Figure 6.6   Learning using multiple learners.

through interactions with the environment and observed parameters. In a real-time systems we need quick response and validation of results so that it can be converted into decisions. Adaptive real-time models are the adaptive models for real-time scenarios.

6.4.1   Dynamic Adaptation and Context-Aware Learning

The dynamic adaptation refers to the continuous sensing of the environment and adaptation to situation to deal with uncertainty in real-life environment. A simple example of content adaptation is depicted in Figure 6.7.

The knowledge base represents the past learning and knowledge built from the learning. The inference machine and the decision machine allows us to select the learning algorithm. The most important aspect of context learning is to understand

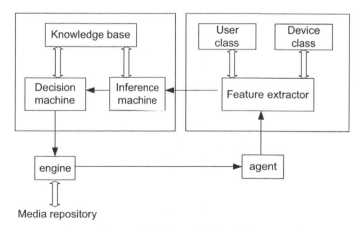

Figure 6.7   Content adaptation.

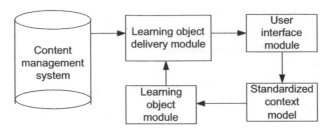

Figure 6.8 Learning architecture with reference to context building.

the learning environment. The dynamic environment demands adaptation; but in the static environment, learning is possible using traditional methods. Dynamic adaptation is possible with the continuous evolution and awareness about present context. This adaptive learning takes place with reference to the context built and helps to build better context further for decision making. A learning architecture with reference to context building and adaptive learning is depicted in Figure 6.8.

There are profile context, learning context, and priority context. While learning, the system should be aware of holistic context when making decisions. The context building with reference to applications is depicted in Figure 6.9. The information captured by different sensors is received by the system. This information comes from multiple sources, and the system builds the context model based on it. The context-aware data fusion helps to combine data from multiple sources to build the overall context. The context is delivered with reference to decision scenario and is used for decision making.

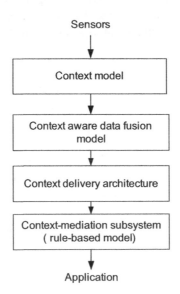

Figure 6.9 Application-based context building.

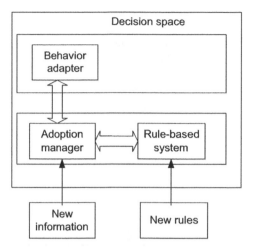

Figure 6.10   Adaptive learning framework.

The important aspect of context-aware environment is proactive interaction with the environment. Knowing the context is a key issue in improving the learning and decision making in the dynamic scenario. The combination of context awareness with dynamic adaptation allows intelligent learning in the dynamic environment. Context awareness is about the awareness of changes in the environment, the relationships and their importance, and the relevance of parameters in decision space. The context is about characterization of a situation of an entity, the environment, and their relationships. A learning system should adapt in advance to a new situation to allow building of the knowledge and reacting to the decision problem in a most appropriate way. Adaptive learning is about knowing the present context, adapting to a decision scenario to create knowledge base, and developing intelligence. Furthermore, this adaptive learning is evolutionary, in the sense the learning parameters and scenarios evolve as we come across new parameters. Adaptive learning evolves as new information, and scenarios are faced. With reference to the above discussion, a framework for adaptive learning is depicted in Figure 6.10. The learning takes place in the decision space and takes the input from the decision space and environment. The behavior of the system is sensed, and the behavior adapter tries to help in the selection of learning policy. The adaptation based on that is done with reference to the knowledge base. The rule-based system or the Bayesian likelihood algorithm can be used for the selection of the learning policy and further selection of learner.

6.5   SYSTEMIC LEARNING AND ADAPTIVE LEARNING

Systemic learning is not possible without adaptive learning. Rather, adaptive learning is one of the important properties and aspects of systemic learning. Since the system consists of various parts and many information sources and agents, it forms a complex

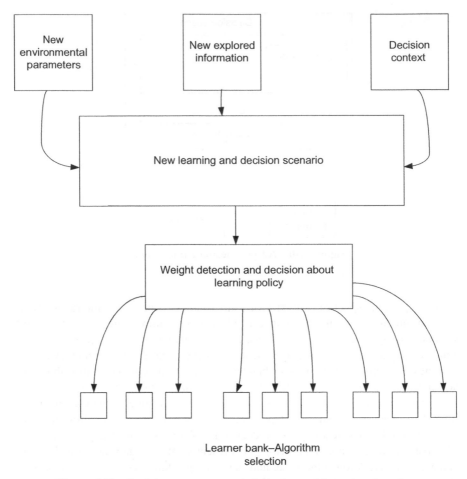

Figure 6.11   Decision system: more information and impact on learning.

environment with multiagents, the distributed information sources, and dynamic changing scenarios. The dynamic environment demands the learner to be adaptive for effective decision making. The changes take place related to the system and decision environment. There are two types of changes: In the first case, more information about the decision scenario and the environment is revealed in due course. In the second case, the environment and parameters in the system are changed due to some other action and the dynamic nature of the system. *Adapting to the overall system behavior to decide the learning policy is adaptive systemic learning.* Adaptive learning is more about understanding the decision and the learning scenario with reference to systemic knowledge and deciding the learning policy. Figure 6.11 depicts the selection and adaptation of a new learning scenario. New environmental parameters, new explored data, and decision context are used to build new learning and

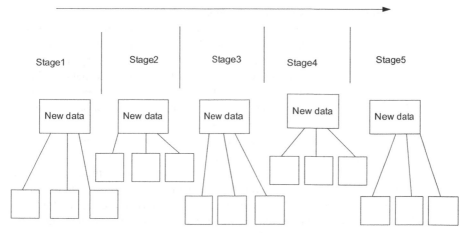

Figure 6.12 Stage-based adaption.

decision scenario. This helps in selection and prioritization of learning policy, and the resultant learning policy is used for learning. The selection of set of learning algorithms and prioritization is done based on the learning policy.

For systemic learning, one may choose to combine multiple learners, since a single learner or learning algorithm may not exhibit the behavior required. Furthermore, a single learning algorithm may not be suitable for all the learning environments. Even in a single learning environment, it may not be suitable for all the learning stages. For adaptive learning, we can combine inductive and analytical learning. The interesting part is to use a combination of learners to get the best out of them and evolving the learning mechanism with reference to learning scenario. Selection of learners and learning policies can be helpful in handling multiple and different decision scenarios.

6.5.1   Use of Multiple Learners

Figure 6.12 depicts the stage-based adaptation. During every learning stage, the learner and system is exposed to new scenarios, new data, and new relationships. The learning policy and algorithms are decided separately during every stage in case of stage-based adaptation. Stage-based adaptation is very useful in case of scenarios where new data becomes available regularly.

The competitive selection of the learner for the decision scenario makes sense when we are using multilearner systems. The easiest and simplest way of using multiple learners is to train different learners to solve the decision problem. This can help in reducing the bias towards a particular learning approach. Use of different training sets for different learners can also help in handling a variety of learning scenarios. The problem in this approach is the unavailability of details about the decision scenario. Another issue is the method that is used to combine the decisions or learning of these different learners. The bank of learners complementing each other can lead to better results. Boosting and cascading can help in optimizing the learning and decision-making performance of the bank of learners. Even in some cases, the training emphasis is given on the data for which the other learners are not performing that well. We will discuss regarding the use of multiple learners for adaptive learning in the next section. Different algorithms used by different learners make different assumptions about the decision scenario and the data and hence can lead to different classifications and decision outcomes. Having more than one learner can make decision making and learning free from the single assumption or predefined fixed assumptions that may not be valid in all decision scenarios. Furthermore, different base learners can be trained for different scenarios with different training sets. These multiple learners work as a set of experts with each of them having their own area of expertise and are contributing to the overall learning process. Moreover, it minimizes the bias and helps in arriving at a fair decision. There can be two types of combination of sets of learners. In one combination, multiple learners work in parallel on the same data. It is a sort of multiexperts combined for learning. All experts or learners give their decisions without taking into account the opinion of other learners. In this case, voting or some sort of weighted average can be used to come to a conclusion. In this approach, a simple weighted sum can be used for final decision making. Here "outcome" is the final decision or learning outcome, while O stands for outcomes of individual learners.

$$\text{Outcome} = \sum_{i=1}^{n} w_i O_i$$

This method does not allow the other learners to take help from any other learner. This method lacks cooperative learning and effective knowledge utilization for learning.

Methods such as bagging and boosting can be used in the process of learning. Bagging was introduced by Breiman [1]. It is derived from bootstrap aggregation, and it is simple and effective for ensemble learning. It can be thought as a special case of model averaging. Along with decision tree it can be applied to different models for classification. The important part of this method is the use of multiple versions of training sets by using bootstraps, that is, a self-initiated process of sampling with replacement. Each of these training sets are built for different models and used to train different models. The final outputs are determined by combining the outputs of different models using averaging or voting. Bagging is effective in the case of unstable models. Hence, it can be viewed as an effective technique in the case of a highly dynamic scenario where building a stable model is difficult.

Boosting is a very popular and widely used ensemble method, which can be used for learning, classification, and regression. This method is based on initially creating a weak classifier, which is not accurate but better than a random guess. Succession models are built iteratively by training based on the dataset, in which the points that are misclassified in a previous model are given higher weight. Finally, all successive models are weighted according to their effectiveness and are measured in terms of success, and then voting or averaging is used to combine outputs to derive the final output. AdaBoost [2] is an

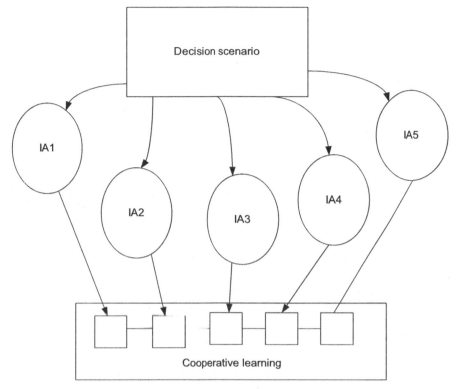

Figure 6.13   Cooperative learning.

adaptive boosting method. This method uses the same training set again and again and can also combine the number of base learners. This makes it possible to get the best out of more than one learner. Here the final outcome is the weighted sum of different outcomes of "n" learners. This approach lacks the use of adaptive intelligence based on the decision scenarios. Also, in this case no two learners interact to improve their own performance. Cooperation and interaction among different learners can help to make the learning process more intelligent and hence can help in dealing with dynamic learning scenarios. Hence, cooperative learning can be suitable for adaptive machine learning. The idea is to use the intermediate outcomes and, instead of weighting, using cooperation among learning processes. Figure 6.13 shows cooperative learning. Here IA1,.., IA5 are intelligent agents learning cooperatively and interacting with one another.

In another approach, the serial method for learning is used. Here, based on increasing complexity, the ambiguous learning scenarios are passed to the next level. This learning takes place in multiple stages.

One more approach is where multiple learners work in parallel and cooperate for decision during every stage. Then again during each stage, learners work in parallel. This cooperation results in learning parameters, weights, and, hence, a better decision. This allows correction of decisions and tuning of learning parameters. This learning scenario is depicted in Figure 6.14.

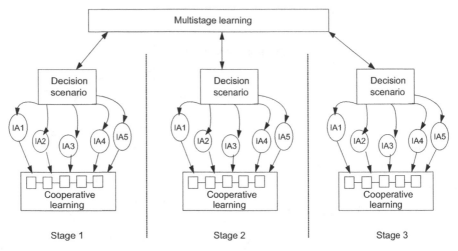

Figure 6.14 Multistage cooperative learning.

Stage 1: $O_{11}, O_{12}, \ldots, O_{1n}$ are the outcomes.
The cooperation among them takes place at the end of stage.

$$O_1 = \sum_{i=1}^{n} w_{1i}O_{1i}$$

Similarly $O_2, O_3, \ldots, O_m$ are calculated and further outcome is calculated.

$$\text{Outcome} = \sum_{i=1}^{n} w_i O_i$$

6.5.2 Systemic Adaptive Machine Learning

The systemic adaptive learning is the learning where learning is adapted with reference to a systemic state. The environment and system parameters are continuously monitored in this case. The system parameters can help to infer the state and stage of the system. The overall system behavior and the set of parameters help in deciding weights for the learner and adapting to the best possible learning policy. Figure 6.15 depicts adaptive systemic learning. Here adaptive learning is based on systemic inputs s_1 to s_{10}. Adaptive learning takes place based on systemic inputs received.

In this case, a model of the system and environment at time "t" is considered for a decision about adaptation at time t. The selected bank of learners is used throughout the learning cycle, and any learning technique can be adapted with reference to the environment and the system. Adapting is based on the system, the user, and the learning patterns. Adaptation can be possible in various ways:

Pattern-based adaptation
Exploration-based adaptation
Forecasting-based adaptation

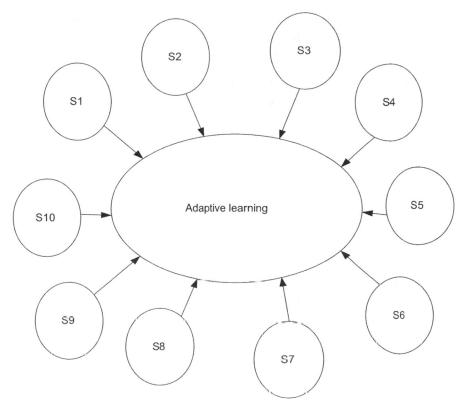

Figure 6.15   Adaptive systemic learning.

In pattern-based adaptation, the pattern or the behavior of new decision scenario is used for selection of learning policies. Based on the behavior of the decision scenario, the appropriate learning policy is selected. In the case of exploration-based adaptation, the new information explored is used for tuning the existing learning policy. In the case of forecasting-based adaptation, the parameters, dependencies, and the context for the future decision scenario are determined based on forecast that is based on historical patterns. The adaptation begins as the change is forecasted. In the case of highly dynamic scenarios and the ability to forecast reasonably, this learning policy can provide better adaptability.

In adaptation there are two types

Static adaptation
Dynamic adaptation

In static adaptation, the selection among the learning policies is done, but the learning policies are static in nature. However, in the case of dynamic adaptation based on changes in the decision scenarios, the dynamic selection and dynamic adaptation of learning policies take place. Here the adaptation refers to adaptation of an overall decision scenario, which includes adaptation of information, relationships, parameters, and dependencies. Let us discuss about how the adaptation occurs with them.

- Adaptation of the information, relationships, and dependencies

    Information is adapted related to the surrounding and scenarios in the context of decision making. The system comes across different scenarios, and the dynamic environment builds new scenarios; adapting to these scenarios in the context of systemic parameters is necessary.

    Adaptation takes place with reference to historical information and present state with reference to historical information (e.g., pattern-based and adaptive learning).

    Adaptation based on dynamic circumstances with reference to new information as it becomes available is equally important.

- Adaptation of the process

    For adaptation of the process, adaptation with respect to the following cases takes place:

    Adaptation of interaction and dependencies with reference to the possible actions.

    Adaptation of the order of the tasks and the steps.

Adaptive systemic learning needs input from the system and environment to decide learning policies.

6.5.2.1 *Advantages of Adaptive Systems*   The adaptive systems can help to make decisions in many complex and dynamic real-life scenarios. There are many advantages of adaptive systems. Some of the important advantages are listed below:

1. Use of relevant information towards learning.
2. May require less number of steps for learning and correcting wrong behavior.
3. Can effectively use the multiple inputs available.
4. Exhibits the required intelligent behavior in a dynamic scenario.

6.5.2.2 *Disadvantages of Adaptive Systems*   Though many advantages are observed in the case of adaptive learning, it may not be suitable for all scenarios and exhibits complexities while training and usage. Some of the disadvantages of adaptive systems are listed below:

1. Bad learning in some scenarios. Adaptive systems may learn wrong moves from bad players in the case of adaptive learning systems for games.
2. Adaptive system-based trainer may find it difficult to train novice and outsmart novice user.

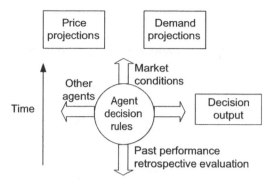

Figure 6.16   Design for adaptive learning.

6.5.3   Designing an Adaptive Application

It is always challenging to design an adaptive learning systems. Multiple learners, different learning policies, and numerous dependencies are features of adaptive systems. Adaptive learning can be designed using multiple learners and a bag of learning policies. In the case of a predefined bag of learning policies, the best one can be selected for decision making.

An adaptive system is a system that is capable of self-adapting and self-learning in response to changing scenarios, inputs, and environments. Adaptive systems should be contrasted with static learning. A static system will not have any self-correcting abilities and will typically behave in the same nonadaptive way until its termination or a forceful interference happens by another system or human. Nonadaptive systems are not equipped for self-modification or correction and are only able to function within a small range of environmental change. Static learning systems are unable to adjust to the new environments, novel environmental scenarios, and some unexpected changes. An adaptive system, on the other hand, will be equipped for self-correction into different system-states in order to navigate, function, and succeed within the different and new environments. It has the ability to respond to the environment. The static systems are introduced with a degree of adaptation. There are always a few functionality constraints and limitations. There is always a degree of adaptability. A typical design for adaptive learning is depicted in Figure 6.16.

The complex environment and dynamic scenarios increase the complexity of adaptive learning. The continuously changing dynamic environment demands adaptive learning in order to improve decision making. A typical framework for complex adaptive behavior is shown in Figure 6.17.

6.5.4   Need of Adaptive Learning and Reasons for Adaptation

Intelligence demands adaptation. The scenarios are dynamic, and without adapting to the new scenario the learning may be incomplete and may not even be able to handle various decision scenarios. Adaptation allows improving achievements and efficiency

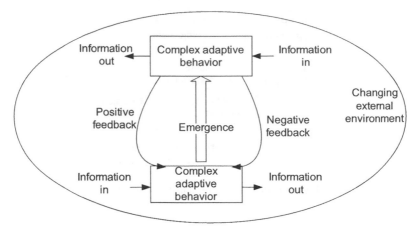

Figure 6.17   Complex adaptive behavior.

in learning with new and unknown scenarios. The past results and successes are used, and their relationships with new scenarios decide the learning policies. Another major advantage of adaptive learning is a decision-scenario-driven and learner-centric environment, which is not tightly coupled to any predefined learning strategy. Furthermore, it gives the flexibility to respond to a decision scenario.

6.5.4.1 *What Can Be Adapted with Reference to Decision Scenario?*
The decision scenario can provide information about the decision environment, the decision objectives, actors, and new parameters in the decision space. Adapting to the decision scenario is understanding the decision environment in order to accommodate explored facts in a learning scenario to improve decision making. The parameters, new patterns, changes in dependencies, and new groups as well as new similarity measures are adapted for learning.

6.5.4.2 *Recommendation Related to Adaptation*   The most interesting part about adaptation is about understanding the new scenario and then relating it to the knowledge base. There is a need to collect the right information and remove outliers. The pattern-based techniques rely too much on past experiences. In the case of adaptation, the dependency and scenario-based outlier removal needs to be used. The information can reveal many new facts in a particular context. These facts need to be used appropriately.

6.5.5 Adaptation Types

As discussed above, the design of adaptive learning process is based on understanding data and information about decision scenario. The adaptation learning types are based on the process of analyzing these data and the reasoning used. We have already discussed different adaptation models. Adaptation types are based on the process of adaptation. The adaptation types used for learning include the following methods.

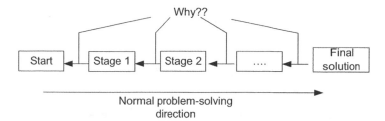

Figure 6.18 Backward reasoning.

6.5.5.1 Forward Reasoning Adaptation is based on user model information. The model in this case tries to adapt based on the forecasted reasoning. In this case, the historical information and experiences are used for forward reasoning. Based on the available information in forward reasoning, what would happen is determined. That is further used for adaptation.

6.5.5.2 Backward Reasoning A decision is made regarding the need of user model information and need of information about the event in rule based adaptation. This is based on the final goal, and why a particular outcome will be there is determined. This is depicted in Figure 6.18.

6.5.5.3 Event-Based Adaptation In the case of event-based adaptation, an event is used for the basis and information for adaptation. Here both forward and backward reasoning can be used. Event-based adaptation does not look for patterns but is more event-centric and useful where every event is indicative of behavior.

6.5.5.4 Pattern-Based Adaptation Pattern-based adaptation is based on information, data, or behavior pattern of the system. The system tries to track the patterns and changes in patterns. With reference to changes in the pattern, the overall learning policy is adapted.

6.5.5.5 Feature-Based Adaptation It is very useful for bootstrapping adaptive applications. Feature-based adaptation allows for many more variations. Each feature is tested for relevance and based on relevance; it is considered and used to adapt something.

6.5.5.6 Uncertainty-Based User Model for Adaptation In case of uncertainty-based user model for adaptation the extent of uncertainty is used to take the adaptation decision and the learning policy selection.

Figure 6.19 depicts relationships among system, user, and adaptation. The adaptation effect is sensed at the outcome and can come in the form of reward or penalty which can further be used for adaptive learning.

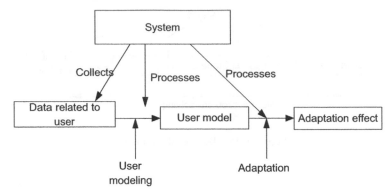

Figure 6.19   Relationship between system, user, and adaptation.

6.5.5.7 *Changing System Boundaries* Adaptation may not be limited to selection of learners, but it can go beyond that. The system boundaries may change with the new scenarios and new decision problems. The adaptation of systems' new boundaries and learning with reference to new scenario is required. The environment keeps changing. The new environment supplies decision parameters, and these parameters help in building the decision scenario. A typical scenario of changing environment in real-life activities is depicted in Figure 6.20.

The decision scenario is also associated with a context. The information may come from a different platform and may be heterogeneous in nature.

6.5.5.8 *Adaptation in Time and Space* The adaptation is a function of changing scenario, and adaptive learning learns with reference to the scenario. The scenario changes over time and space, and the new information becomes available in time and space dimensions. The relationship between this new information and

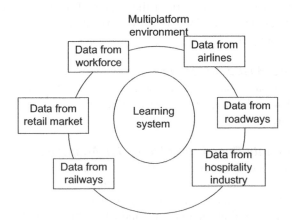

Figure 6.20   Multiplatform environment.

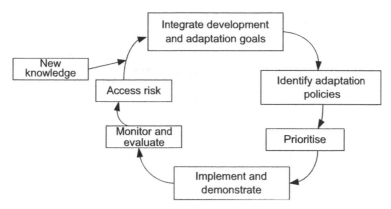

Figure 6.21   Adaptive framework: a continuous process.

reference to the decision scenario plays a role in adaptation. Adaptation needs to consider the learning with reference to time and space and exploration with reference to these parameters in the provided decision scenario.

6.5.6 Adaptation Framework

Adaptation framework gets input about new scenarios, new information, and parameters and adapts to new scenario and formulates a learning policy for new information. The framework should understand the impact of the new information on dependencies and relationships and should formulate or select appropriate learning policy for the information. The information made available from different sources helps to identify the relationships and changes in the decision impact. This further helps to build the decision context. With reference to knowledge base, a decision scenario is built. The adaptive framework depicted in Figure 6.21 uses this decision scenario to decide the learning policy and adapt to a new decision scenario. It is a continuous process and the best learning policy with reference to decision scenario is adapted.

The framework discussed here will help in the selection of an adaptation strategy or give the guidelines for the implementation of adaptation actions. There are a number of necessary and important stages in developing an adaptation and learning strategy. Adaptive learning is an ongoing process of engagement with the changing environment and decision scenarios. It consists of following stages and has a number of key elements.

- An iterative process for selection and checking the decision parameters. It is required so that any strategy is constantly updated based on relevant decision parameters.
- Exploration of new actions. The exploration of new actions and impact of those actions helps in adaptation.

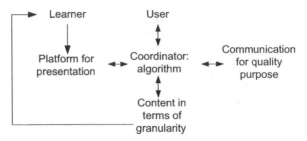

Figure 6.22   Adaptation framework.

- A mechanism to sense information such as relevant decision parameters and understanding the relevance of those parameters.
- An adaptation function. The adaptation function is derived to adapt to a new decision scenario.

Figure 6.22 depicts an adaptation framework typically to maintain the decision quality. It receives feedback from the environment regarding the quality. The coordinator algorithm is responsible for coordination among user expectations, responses, and parameters and granularity. The learner adapts to them and decides the learning platform and learning policies.

6.6   COMPETITIVE LEARNING AND ADAPTIVE LEARNING

Adaptive learning can be competitive. In competitive adaptive learning, more than one learner contributes to adapt. Concept adaptive rival penalized competitive learning (RPCL) was introduced by Cheung et al. [3] This includes training and prediction stages. Here, sliding window scans through input–output pairs. In this approach, predictors adaptively learn in the prediction stage. All the information is not available at a particular instance, and the input–output relationships are made available in particular time slots. These may be available in the form of snapshots taken during the time interval. Sometimes two independent time slots cannot be used effectively. To make the best use of relationships and establish relationships, a sliding-window mechanism can be used. The sliding-window mechanisms with overlapping time stamps are depicted in Figure 6.23.

In competitive adaptive learning, the learners compete to meet the predicted requirements of the system. In this case, one or more learners fit for the decision scenario or the ones suitable for the predicted scenario are employed for learning. A typical architecture for competitive adaptive machine learning is shown in Figure 6.24. In this figure, I_1 to I_8 are input sources and information. This information is used to predict the decision scenario, the learners L_1 to L_5 compete with reference to the decision scenario to achieve the outcome, and the learning policy is built as a result of competition.

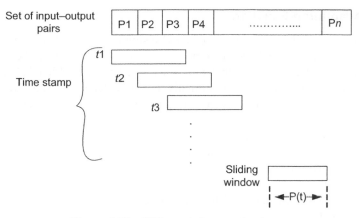

Figure 6.23   Sliding-window mechanisms.

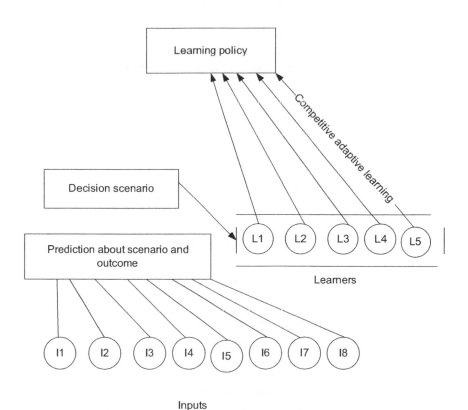

Inputs

Figure 6.24   Competitive adaptive machine learning.

6.6.1 Adaptation Function

The adaptation function is dependent on the decision scenario and parameters. There is a need to map the decision scenario to learning policy and the best learning algorithm. The adaptation function does this task. It gives the likelihood of predefined learning policy that it would learn optimally. It is time dependent and gives the best possible policy at a particular instant. Representation of the parameters and the function is discussed.

The environment parameters: $\{e_1, e_2, e_3, \ldots, e_n\}$

The forecasted environment parameters: $\{fe_1, fe_2, fe_3, \ldots, fe_n\}$

The error $= fe_1 - e_1$

Adaptation function $= F(\text{error})$

New decision parameters = f(new scenario, new parameters, old parameters, adaptation function).

6.6.1.1 *Decision Parameters and Scenario* Here we will discuss decision-centric active learning with emphasis on the decision parameters and decision scenarios. One of the major parts of active learning is to get labels for all unlabeled samples. This should proceed in such a way that it should produce a better model. The decision alternatives are considered while learning and decision making. The decision parameters refer to parameters that are relevant parameters and influence the decision-making process. Decision-theoretic models are based on mathematical/statistical decision theory concepts. A general knowledge based decision making is depicted in Figure 6.25.

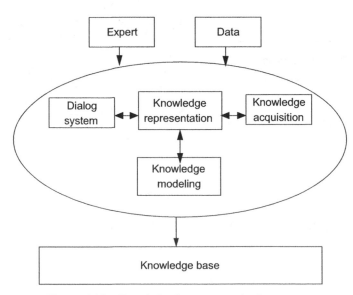

Figure 6.25 Knowledge base system: basic structure.

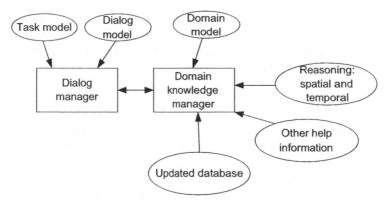

Figure 6.26 Domain knowledge and dialog manager: basic view.

In the case of decision-centric learning, the idea is to acquire decision parameters, and based on that the explored facts and unlabeled information is used judiciously. Here learning is based on decisions and its impacts. In the case of decision learning, the optimal decision is used to empower learning and classification. In the case of response learning, the predictive model estimates the probability of action outcomes, which is used for learning. The decision scenario leads to different decision parameters. Domain knowledge, historical information, explored decision parameters, and decision scenario are used for deciding the learning policy. The basic architecture of domain manager along with the dialog manager is shown in Figure 6.26. Domain knowledge is required to build the context. The domain knowledge and decision interactions are basic inputs for domain manager.

The selection of decision parameters is based on the historical pattern and the likelihood of impact of the parameter on outcome. This is generally based on the relevance of the decision parameter. The various statistical methods can be used for selection of decision parameters for a given decision scenario.

Decision parameters are used to develop two mathematical models for the decision-centric learning. The ranking of the variables decides weights of the parameters in learning and decision making. The decision network is built using data inputs, expert inputs, different parameters, and utilities. The decisions are refined with all these inputs. A typical decision network is depicted in Figure 6.27.

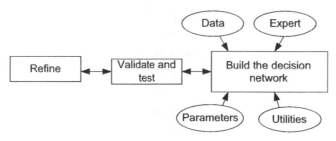

Figure 6.27 Building the decision network.

For adaptive learning we need to manage decision information, relationships, and attributes. The prioritization of attributes with reference to the decision scenario is required. Further evaluation of information in different contexts across different levels is followed by adaptation of learning policy.

6.6.2 Decision Network

In the decision network, a decision evolves based not only on information but also on interpretation of that information. Another factor is the representation of environment. A decision network includes information about the agents' present state, possible actions, possible outcomes, and transitions. Based on some evidence or information about the environment, a particular action is taken.

Let A be the action with reference to knowledge instance KI. Then the expected impact (EI) of action A can be given by Bayesian

$$EI(A/KI) = \sum_i Outcome_i$$

A decision network can deal with multiple actions and outcomes. A decision network includes events, dependencies, and impact along with decision parameters. Different decision attributes with reference to adaptive learning are depicted in Figure 6.28. The filtered data are interpreted, and it helps to build perceptions. The decision maker uses these perceptions, while different attributes are used to generate intentions. A set of behavior routines for references are used by the decision maker for decision making about parameter control.

6.6.2.1 *Decision-Based and Problem (Scenario)-Based Learning* In the case of learning based on historical data or predefined learn sets, the information available at disposal is used for learning. This has limitations such as failing to capture

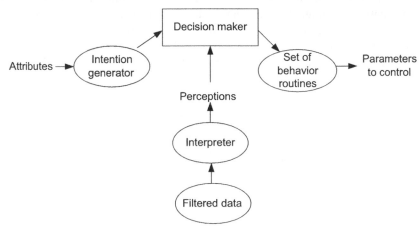

Figure 6.28 Working of decision maker.

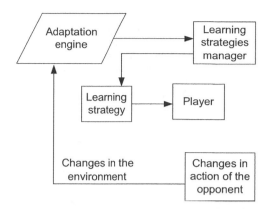

Figure 6.29 Adaptive learning: change with the environment.

new scenarios and making wrong decisions in case of unknown scenarios. But in the case of exploration-based learning, these decisions and the outcome can be helpful for learning. Decision-based learning tries to exploit the information about the decision, the decision scenario, and impact of the decision making while learning. All learning pointers come through the historical decision making.

In problem-based or scenario-based learning, the perceived relevance is important. Furthermore, as the scenario unfolds, there is a need for cumulative learning. Problem-based learning demands active learning. With new information the old assumptions that are no longer valid are corrected. In another option, a scenario is presented for learning with different possible parameters. It is more like simulated exploration. The most appropriate, relevant, and frequent scenarios help in providing better knowledge building.

6.6.3 Representation of Adaptive Learning Scenario

Let us assume that a boxer is playing where the assumption is that the opponent is a left hander and hence will attack with the left hand. The player will have his defense planned as per this assumption, and even attack is planned as per this assumption. Soon the player realizes that the opponent can punch equally well with both hands— now he has to adapt to this environment to win the match. These adaptations are based on body language of the opponent, his responses to different actions, and the outcomes in due process. A typical adaptive learning scenario with respect to the example discussed is depicted in Figure 6.29.

As discussed so far, the learning strategy is major outcome of adaptive learning. It is based on knowledge acquisition, abilities, and learning patterns. The aims and challenges with reference to learning strategy are depicted in Figure 6.30.

In the case of learning, a teacher selects learning strategy by adapting to the student's interests and strength. These inputs can be sought through questionnaire or interaction with the students. Based on that the course content can be designed, the way of delivery can be improved, and even teaching policy can be finalized. Here the

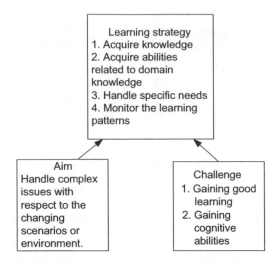

Figure 6.30 Learning strategy: aim and challenge.

parameters include interest of the students, their academic background, other courses, and the overall objective of the institute.

6.6.3.1 Complementing and Selecting Method for Adaptive Learning
While selecting machine learning models in the case of adaptive machine learning there is a need that models should complement each other. The basic method needs to be selected based on its appropriateness. To address the varied population and different scenarios, a bag with complementing techniques can be helpful. This may typically include a mix of exploitation and exploration techniques, quantitative and qualitative techniques, knowledge-centric and inference techniques, sequential and parallel-learning techniques, and voting and filtering techniques. Complementary method selection can offer the comprehensive learning required to handle complex decision making.

6.6.3.2 Complex Learning Problems and Need of Different Learning Methods
The learning problem in real life needs to deal with complexities. The same learning method cannot deal with different decision scenarios. Hence to deal with complex decision problems we need different learning methods.

6.7 EXAMPLES

Different adaptive learning classification problems are discussed in this section.

1. *Document Classification*: With information explosion—there are huge document sets that need to be classified. These documents include research papers, news articles, business forms, personal documents, and bank documents. Even in the case of a particular domain, there are many different documents that need to be

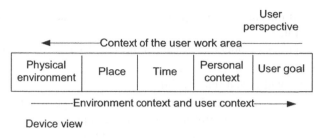

Figure 6.31   Environment context and user context.

classified. The availability of new documents and new decision problems demand a change in learning policies. When the same set of loan documents are to be classified for underwriting versus for property evaluation, there is a need to adapt the decision scenario to create the right buckets.

Other adaptive classification applications include the following:

1. Satellite image classification
2. Student behavior classification

6.7.1   Case Study: Text-Based Adaptive Learning

Text-based learning needs to understand context, and context is closely associated with a decision scenario. Various methods and solutions such as SVM, boosting, and K-nearest neighbor can be used for text classification of text. Learning with adaptation of the environment and the decision scenario is required. The adaptive learning with reference to new information helps in building a knowledge base. Furthermore, new information can change the decision context. There is a need for enhancement in the behavior of a learning system with reference to changes in environment. In simple adaptation there is a change in fixed rule. This adaptation can be taken to a higher level of flexibility to exhibit intelligent behavior. The information can be adapted to user, circumstances, and actions. Adaptation can be in the form of order of tasks or change in the behavior or process of interactions. Here the system can adapt to goal, decision parameters, and user inputs. For example, for a person with research background searching for machine learning papers, the system can adapt to these inputs. It can adapt to his previous searches and old queries, along with the recent scientific advancement to learn so that the best possible outcome can be produced. The relationship between the environment context and user context is depicted in Figure 6.31.

The intelligent adaptation can include

- Inserting/removing information

    Prerequisite explanations and context-based information for document mining can be inserted by adapting to user needs.

Additional information or explanation can be provided by adapting to user behavior.

A relevant set of documents can be provided based on the overall user behavior.

- Altering information and relationships

  With reference to new actions, information can be provided or altered.
  Adapting to social and information context can provide recent information.

- Adaptive classification of documents/text. Establishing new relationships and group formation to adapt to user needs.

- Providing additional information by adapting to information patterns.

- Adapting to environmental changes with reference to complete repository of knowledge.

- With reference to adaptation of user behavior do not provide some information.

In this way the most relevant information can be provided to the user, and text classification can be performed in an adaptive way. Even adaptive navigation support can be provided. So for research paper selection, context may include background of user, his previous searches, the requirement of projects, his colleagues, his research labs, and soon.

6.7.2 Adaptive Learning for Document Mining

There are huge sets of documents, and getting the most relevant documents from the heap of documents remains a challenge. Every time we want a document for different objective. The information needed may have exact context and similar information that may not be useful. The mining in this case needs to adapt to the scenario of decision making. The adaptive document mining can be a solution for this. There is a need to represent the objective, and the objective can be tracked based on behavior of user. The behavior of documents and role in various decision processes need to be considered. Even the user behavior can help in adapting. *Adaptive behavior-based learning* is learning based on the adapted behavior of the system and the observed impact on results. *Adaptive document presentation* is with reference to the customer behavior and the problem at hand; the documents relevant to the decision scenario are presented to the user. Furthermore, *adaptive regrouping and adaptive classification* of document is possible using adaptive learning.

Initially, the user is interested in sports document and mining them and not interested in political documents that also talk about sports. The decision scenario here is to identify the recent sport activities. The decision scenario changes when the user now is a sports journalist who is interested in news about inauguration of a sport event. In the new decision scenario, all sports events related to inauguration of a sport event are grouped into a cluster through adaptive learning forming a different cluster structure. Furthermore, adaptive learning can help in adaptive navigation of documents where document classification is adapted with reference to the decision scenario. This will allow adaptive knowledge building and representation.

6.8 SUMMARY

The changing scenarios, dynamic environment, and uncertainty in real-life decision problems pose various challenges for machine learning in real-life problems. Adaptive learning can help in many scenarios to lead to better learning and decision making. Adaptive learning is about adjusting with the changing environment and context. Since the behavior is changing, the system comes across new information and new decision scenarios that may offer completely new and different behavioral information. The learning policy that was fine so far may not be relevant in the new scenario. The adaptive learning can adapt to decision scenarios, new knowledge, and user behavior and user inputs. An *adaptive system* is a set of various entities, which are independent or interdependent, real or abstract, forming an integrated whole that together are able to respond to environmental changes or changes in the interacting parts.

Selection of learning algorithms or learner and classifier from the existing set of classifiers based on a decision scenario is one way of attacking this problem. Since this approach has predefined classifiers, there is a need that these classifiers should evolve or adapt to changing behavior to exhibit truly learning and intelligent behavior. Cooperative multistage machine learning allows us to adapt in different phases, and the learning is cooperative; hence, the overall learning is benefited from different learning agents. Adaptation needs to capture the decision scenario and relevant parameters. The decision scenario and relevant parameters are used in adaptive learning. Adaptive learning is an important part of systemic learning as the systemic dependencies reveal as the learning progresses and the system needs to adapt to the environment.

REFERENCES

1. Breiman L. Bagging predictors. *Machine Learning,* 1996, **24**(2), 123–140.
2. Freund Y and Schapire R. Experiments with a new boosting algorithm. *Proceedings of the Thirteenth International Conference in Machine Learning,* 1996, 148–156.
3. Cheung Y, Leung W, and Xu L. Adaptive rival penalized competitive learning and combined linear predictor model for financial forecast and investment. *Proceedings of IEEE/IAFE Conference of Computational Intelligence for Financial Engineering;* 1997.

CHAPTER 7

Multiperspective and Whole-System Learning

7.1 INTRODUCTION

As we discussed so far, there is a need to use all resources, information, data points, and information sources effectively and optimally while learning. Hence all direct and indirect knowledge sources should be utilized. When we consider whole-system learning, the expectation is that the details about the complete system with reference to possible actions should be learned. Whole-system learning is a structured approach to engage learners' head, heart, and hands, precisely the learner's "whole system." Similarly the concept of whole-system machine learning (WSML) involves engaging all the available information and resources to the best of their capacity for learning. This refers to engaging all information sources, percepts, and action points. One of the important aspects of whole-system learning is to capture completely the system information and knowledge, which needs to be captured from different perspectives. Interestingly, multiperspective learning is mandatory for whole-system learning.

Since more and more information is available and accessible, it becomes even more important to use this information for learning optimally. Each part of the information offers a certain perspective. Some perspectives are very important for the decision scenario, while some of them are not so important. In the absence of knowledge about different possible perspectives, decision making becomes even more difficult and learning may remain incomplete. Engaging the whole system here refers to engaging all possible information avenues for effective learning.

Perspective is a viewpoint or rather data gathered and arranged with a few assumptions. These assumptions define the viewpoints of looking at the system and actions. Information can be gathered, processed, and presented from different perspectives resulting in different decision drivers. When more than one intelligent

Reinforcement and Systemic Machine Learning for Decision Making, First Edition. Parag Kulkarni.
© 2012 by the Institute of Electrical and Electronics Engineers, Inc.
Published 2012 by John Wiley & Sons, Inc.

agent is involved in the learning process and each of these agents is collecting information from a particular perspective, then all these agents together can make whole-system learning possible. The multiperspective learning increases overall learning complexity but introduces many more learning opportunities compared to traditional learning mechanisms. The information assimilated by different agents working with different assumptions provides the picture of the system from that particular viewpoint or perspective. The interactive and active learning allows considering these different perspectives and helps us to combine them for the overall perspective to learn.

Whole-system learning is also multisensory learning where the information is coming from heterogeneous intelligent agents. The information comes from different sources and is in different forms. The effective learning has the following properties: It is contextual, it is incremental and cumulative, it possess the ability to provide integrated view, and it also has to be proactive, collaborative, and reflective. All these properties make learning more and more complex, but it can address some key issues in the case of complex learning scenarios. Whole-system learning is very important from the perspective of getting the complete insight into the system. Multiperspective learning is the process of using data and information collected from different perspectives,—that is, parameters, values, and objectives for learning—so that the diverse perspectives are considered for decision making. Perspectives are also defined at the level of assumption. When there is more than one information source and the information is coming from different perspectives, it is highly likely that this information is heterogeneous. There is a need to bring the different information and knowledge representation on a common platform.

In short, multiperspective machine learning (MPML) and WSML are two important aspects of systemic machine learning. While WSML tries to use the all the information and aspects effectively for learning, the MPML tries to get all the information required for whole-system learning and tries to integrate this information. Figure 7.1 depicts the role of MPML and WSML in systemic machine learning.

7.2   MULTIPERSPECTIVE CONTEXT BUILDING

The context here defines understanding the state of the system when learning is taking place. When decision needs to be taken based on learning, we need to understand decision context. So there is a need to capture and represent different perspectives. The most important part is about understanding different perspectives. Any object, scenario, or event has many perspectives. The decision context can only be described by understanding all these perspectives. Context is a sort of collective knowledge giving ins and outs about the decision scenario. The information about scenario builds the context. This includes the decision objective, the environment, different parameters, and most importantly the relationships and dependencies among actions. The individual parameters in isolation may convey different decision objectives while all parameters together build a decision context. For example, while

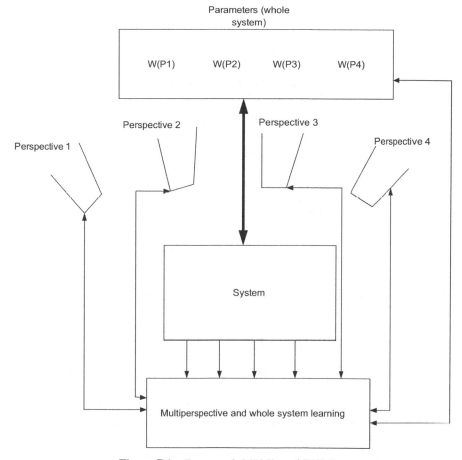

Figure 7.1 Framework MPML and WSML.

searching some research technique the context can be the application, the user, and the present scenario. For different people, implementing this algorithm may have different context. This context can be built through interactions among different objects in the system. Figure 7.2 describes context building through interactions.

The decision artifacts are typically the decision parameters that become available through different sources. These artifacts help in building the context for decision making. Furthermore, this context with the scope and other information helps to determine the decision scenario. The perspectives can typically be opinions of different experts or information available at different decision points in decision space.

A multiperspective intelligent framework for machine learning is depicted in Figure 7.3. With reference to the environment, the perspective is determined. The data acquisition and parameter determination are done from that perspective. With reference to learning and decision strategies the implications of decisions and learning are determined.

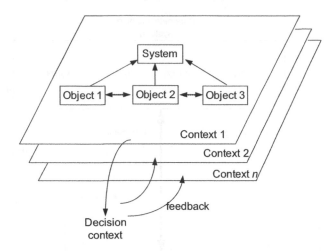

Figure 7.2 Context building.

7.3 MULTIPERSPECTIVE DECISION MAKING AND MULTIPERSPECTIVE LEARNING

As discussed in Chapter 2, multiperspective learning is required for multiperspective decision making. Here multiperspective learning refers to the learning from knowledge and information acquired and built from different perspectives. The multiperspective learning needs to capture information, relationships, and system parameters from different possible perspectives. This process includes methods to capture perspectives and representing and relating captured data, information, and knowledge with reference to the different perspectives. The perspective refers to the context, scenarios, and situations that impact the way we look at a particular decision problem. An intelligent agent captures a sequence of percepts. Here the assumption is that each percept captures all features. In this case, there can be more than one intelligent agent. Each of intelligent agents captures multiple percepts. It also captures

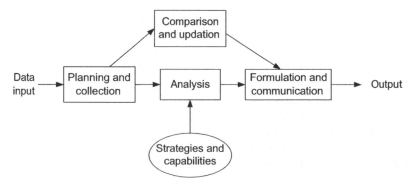

Figure 7.3 Multiperspective intelligent framework.

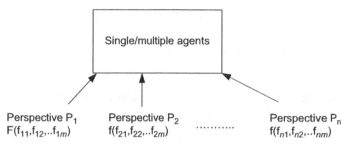

Figure 7.4   Multiperspective learning.

different perspectives and may provide different features. These sequences are separated in timescale, and different agents look at different parts of the system. Multiple agents can capture percepts separated in feature space.

In Figure 7.4, P_1, P_2 ... P_n represent different perspectives. Each of these perspectives is represented as a function of the features and system dimensions. These perspectives are related to each other in terms of features. These features are related and may overlap with features from other perspectives. Two perspectives may share some common part of the system. In some cases, the features are identical but the relationships and weights may change and hence representative values may be different. This difference exists as some part of the system, and properties of the system visible from one perspective may not be visible from the other perspective—or may be partially visible from a different perspective. As discussed previously, the representative feature set should contain all the possible features.

As per the definition; perspective is a state of one's ideas, the facts known to one, etc., and it has a meaningful interrelationship. It is about seeing all the relevant data for a particular problem space in a meaningful relationship from the available window.

7.3.1 Combining Perspectives

As discussed, different perspectives are relevant from a particular decision scenario, and to make multiperspective learning possible there is a need to combine the different perspectives. It is the most challenging part of multiperspective learning. Different approaches can be used for combining multiple perspectives. One simple way is prioritization of perspectives and then combining features based on priority and combining different features. Another approach is that the weighted sum of feature vectors obtained form different perspectives. The representative relationships cannot be generic and always in association with a certain decision scenario. The perspectives can be combined with reference to a particular decision scenario. In another effective technique, more than one perspective is combined with reference to decision scenario. This is done using the prioritization of perspectives and features with reference to decision scenarios. The ranking of perspective is usually performed with reference to decision scenario. Further weight for each of the perspectives and features is determined based on this prioritization. The combination of different perspectives is done using a weight matrix for the given decision scenario. The partial information and its influence on decision should be represented and mapped. This can be achieved through graphical representations and dependency matrix. These representations are discussed in this section.

7.3.2 Influence Diagram and Partial Decision Scenario Representation Diagram

The perspective-based information can be represented in different ways. As there exists a conditional relationship, it can be modeled using Bayesian likelihood or with the use of other statistical techniques. It can be represented as an influence diagram (ID) or conditional relationship diagrams. This ID-based representation can help us to identify relationships and determine the context. ID is a graphical representation of decision (scenario) situation. There can be other ways to represent decision situations and relationships. We have chosen ID because it can help us most appropriately in representing system relationships, and also it is very simple and a less complicated way of representation.

In traditional learning, the information is generally presented from a particular perspective. But in real-life scenarios there are different complexities and inter-dependencies. Even for a simple problem, there are many possible perspectives. Some of the perspectives are directly related to objectives, while others may have indirect relationships with objectives. Perspectives that can be directly derived from objectives play the major role in analytical thinking and analytical decision making. The multiperspective learning should consider different perspectives. Another important aspect is the decision-making perspective. The decision-making perspective needs to be mapped to the learning perspective.

The fundamental idea of multiperspective learning is to capture systemic information from all possible perspectives. This can help in building the overall systemic perspective. The information from the various perspectives is used to build the context and systemic knowledge, and that knowledge is used for effective decision making.

The ID, decision diagrams, and decision trees are used to represent different kinds of information. The ID shows dependencies among the variables very clearly. In semiconstrained ID the possibility of dependency is shown. Some examples of ID with perfect information, imperfect, and no information are discussed in Chapter 2.

Actually, while learning in real life, cases having all the information at the time of decision making are not possible; hence the imperfect information based case is the obvious scenario. There is a need to model and acquire systemic information in the case of imperfect information. A typical example of imperfect information is depicted in Figure 7.5.

IDs are particularly helpful in the following cases:

- when problems have a high degree of conditional independence,
- when compact representation of extremely large models is needed,
- when communication of the probabilistic relationships is important, or
- when the analysis requires extensive Bayesian updating.

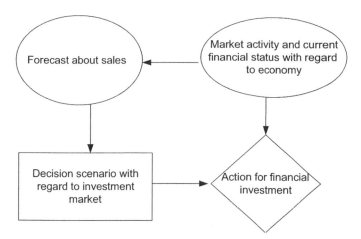

Figure 7.5   Imperfect information (real-life scenario).

The conditional independence allows representing conditional probability in a more useful way for decision-making problems and hence is very important for machine learning. IDs represent the relationships between variables. These relationships are important because they reflect the analyst's or the decision-maker's view of the system.

In short, probability ID is a network with directed graphs but no directed cycles. The ID represents relationships in terms of features. These features may also facilitate knowledge acquisition. The relationship among features helps to build contextual information.

It represents the overall decision scenario. For example, a clinician expert may be able to assess the prevalence of a disease and sensitivity and specificity of a diagnostic test more easily than she could assess the post-test probability of disease. After the ID is drawn to facilitate probability assessments, all updating and Bayesian inference are handled automatically by the evaluation algorithms. Although there are approaches for performing Bayesian updating within a decision tree, for problems with extensive Bayesian updating, such as sequential-testing decisions, IDs ease the burden on the analyst. This is done by reducing the need for complex equations required for Bayesian updating in the tree. IDs also reduce the time required to find errors that may be introduced when these equations are specified.

Here we will use ID for the representation of the decision scenario. In a real scenario, ID is the representation of the part of the system that is visible to the decision maker. We can refer to this as perceived decision boundaries. Also, it can be a system representation from a particular perspective. In real life, it is always possible that even the complete information from the obvious perspective or decision-maker's perspective is also not available at the time of making decision. This limited information about dependency as well as insufficient information for decision making can lead us to represent decision scenarios in a slightly different way; we will refer to it as a semiconstrained influence diagram (SCID), also called a partial decision scenario representation diagram (PDSRD).

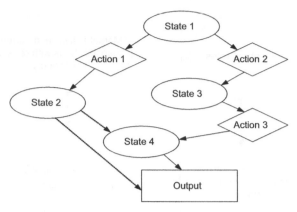

Figure 7.6   Partial decision scenario representation diagram—PDSRD.

PDSRD has the relationships represented in a fuzzy manner. These PDSRD fuzzy relationships become concrete as we combine more and more perspectives and the revelation of systemic information over the period of time. Figure 7.6 represents PDSRD.

PDSRD as discussed represents the partial information and fragmented information from a particular perspective. PDSRD can be further modified with a few changes to represent links where the clear relationship is not available. PDSRD with a probable state is depicted in Figure 7.7.

Dotted lines represented in the diagram indicate the possible relationship. In PDSRD, there can be some relationships you are not very sure about. Some of the relationships are fuzzy and represented with the question marks on the link. The probabilities of transition are known for a few relationships and not known for others. This helps in forming a partially filled decision matrix with fuzzy values.

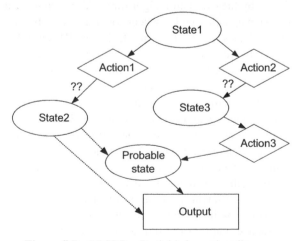

Figure 7.7   PDSRD—Partial information diagram.

As we discussed above, the complete information may not be available in different scenarios. The representation of relationships and mathematical formulation in the case of complete information and partial information are discussed in the next section. Furthermore, one or more perspectives are needed to be accommodated in representation and mathematical formulation to represent the decision scenario.

7.3.2.1 *Representation Based on Complete Information*   Here complete information means that all the systemic parameters are available. This helps us to determine all probabilities of transition. Due to this availability, the decision making becomes much easier. But in real life the complete information is not available at any point of time. When we have a complete picture available or we are absolutely sure that a particular event or pattern-based decision making can very well be used to solve the problem, this representation is used. Actually, ID with complete information is a special case of that of PDSRD. Hence PDSRD is diagram where information availability varies from zero to complete.

7.3.2.2 *Representation Based on Partial Information*   Generally, there is only partial information available. This partial information can be represented using PDSRD. There are many such diagrams from different perspectives, and the ones in isolation cannot guide us to decisions. The representative diagram for all these representations is required for it. The Representative Decision Scenario Diagram (RDSD) is the representation of decision scenarios by combining different PDSRDs. The RDSD is a representation of multiperspective learning. It is actually a representation of the knowledge acquired from all perspectives.

7.3.2.3 *Uniperspective Decision Scenario Diagram*   The PDSRDs that are generally used to represent decision scenarios are uniperspective ID. The transitions and the probabilities associated with those transitions in this diagram represent the decision-maker's perspective. Even ID with probabilities can be viewed as uniperspective decision scenario diagrams (DSDs).

7.3.2.4 *Dual-Perspective Decision Scenario Diagrams*   To overcome the limitations of uniperspective DSD, we represent information in a dual-perspective DSD. Here in a single diagram, there are two probabilities and transition patterns based on perspectives that are represented. The dual perspective IDs can help in representing some of the not-so-complex problems where two perspectives can cover most of the system and decision space. Still it can deal with the majority of applications and can represent the majority of the decision problems.

7.3.2.5 *Multiperspective Representative Decision Scenario Diagrams*
As there are many possible perspectives in real-life complex problems, a decision needs to be made after taking into account these perspectives. Hence there is a need for multiperspective decision problem representation and solving the same. As discussed in the previous section, PDSRD represents different perspectives where a single PDSRD represents a particular perspective. There are PDSRDs for each perspective.

These PDSRDs are used to form an RDSD for a particular decision scenario. The RDSD is used for decision making and allows multiperspective decision making. This is typically a representative diagram for all PDSRDs.

In the case of absence of knowledge of dependency from a particular perspective, the RDSD will not represent that particular perspective. More and more information with reference to perspectives is incorporated in RDSD. Hence DSD represents the best view of decision scenario.

7.3.3 Representative Decision Scenario Diagram (RDSD)

The different PDSRDs represent different perspectives and relationships among features from that perspective. For any system or even for a decision problem there can be many PDSRDs. Each of them represents a particular perspective. Any decision scenario demands information about the number of features for decision making and learning. The relevance of these features gives the weight for these features. All these features are not available in one perspective. But a representative decision diagram combines all PDSRDs with reference to this decision scenario and gives the best possible values for all the relevant features.

7.3.4 Example: PDSRD Representations for City Information Captured from Different Perspectives

Let us discuss an example where different perspectives about the city information are possible from different agencies, PDSRD1 to PDSRD3 representing it.

7.3.4.1 PDSRD1   It provides the travel-agent perspective towards a city and their trips. It also includes the information about the region surrounding the city. It also includes information about facilities for traveling and hotels in the city.

Information about: Traveling arrangement, tourist spots, cost, markets, taxi services, and locations close by.

7.3.4.2 PDSRD2   It provides the perspective of some members of sociopolitical organization who are actively involved in these activities. It provides the politicians perspective towards a city. The information about the people and the other social and political aspects of the city along with the region is also provided.

Information about: People, the political background, religion, communities, social environment, social aspects, political importance.

7.3.4.3 PDSRD3   It provides perspective of cultural experts and historians. It includes cultural, that is historian's perspective towards the city. So, it includes the information about monuments and about the historical importance of the city. Information about cultural aspects of the city is also included.

Information about: Historical monuments, cultural aspects, historical importance of the place.

Consider the following decision scenario:

Decision Scenario: A group of people want to make a decision whether to visit this city or part of the country, considering that they are interested in food and are looking to study the historical aspects of the country.

Representative DSD: This will combine the three perspectives with reference to the decision scenario where the information from historians and travel agents will carry higher weight. Furthermore, each of these perspectives will give different aspects of the city, and the combination of these aspects will build the representative DSD.

Representative DSD will contain information about: {people, travel arrangements, political background, leaders, facilities, hotels, taxi services, tourist spots, parks, places to visit, historical monuments}.

The priorities of the features will depend on decision scenario. If the information is coming from more than one source, then even the source will contribute towards the weight associated with a particular feature. A similar concept of cumulative learning is depicted in Figure 7.8. Here the context is built through different perspectives such as capabilities, subject knowledge, theory concepts, and core competencies. The overall context and all these parameters are used for decision making.

A typical multiperspective view for a learning system is depicted in Figure 7.9. In a learning system there are student perspectives, parent perspectives, teachers' perspectives, and perspectives of other educational officials. The multiperspective learning takes into account all these perspectives. While doing this the inputs from different sources such as technical systems, personal systems, and organizational systems are considered.

Whole-system learning makes use of information coming from all the sources. Whole-system learning in social/organization context is depicted in Figure 7.10. This can even be mapped to WSML. Information here stands for previously learned patterns and data stored in databases. The information exchange allows the combination of information coming from different sources to build knowledge. The information is used for the forecasts and predictions. The actions are used for exploration, and the rewards and penalties are used for learning.

The information comes in various forms and from different sources. The information can be *data stored* from historical experiences (various intelligent agents capture information available to them), *visualization* (visible or directly derivable information), and *action* (the information coming through the exploration). All these activities and sources make learning possible. There is an iterative, cooperative, creative, and

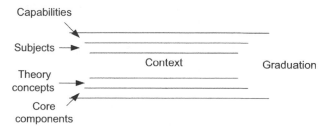

Figure 7.8   Cumulative learning.

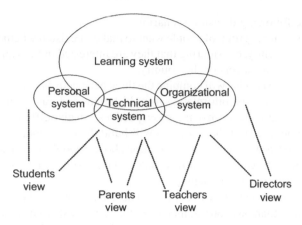

Figure 7.9 Multiperspective view of a learning system.

collective learning cycle. Here's how it goes: (1) Information is captured from the environment, system, and world into different forms. (2) Scenarios or decision problems about specific concerns or questions are used to form decision context. Here the diverse perspectives and information are shared by different parts of a whole system. Intelligent agents with reference to decision scenarios perceive them. Out of that information exchange and mapping comes new information, understandings, relationships, and possibilities. This is acquired in the context of a greater sense of system and environment among the diverse components, intelligent agents, and their perspectives. (3) They are inspired to create a new collective context or perspective together that includes what is most important to all of them with reference to a decision scenario, and (4) they decide to weight the perspectives, prioritize views, and so on, to realize the complete systemic context of taking real action in the world that addresses the problem at hand with reference to a decision scenario. As agent's exploration goes

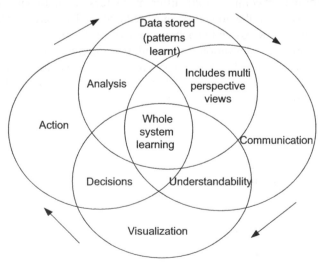

Figure 7.10 Whole-system learning in social context.

on and they experience what actually happens as an impact of their decisions and actions, they review those outcomes and learn in the form of rewards or penalties, and this becomes a new data point for learning and information to be preserved in knowledge base and to be used in similar decision scenarios in the future. Out of that decision framework, learning policies and information, new questions, and allied decision scenario arise and come together from an intelligent agent to overcome flaws of previous learning. Thus the cycle continues adaptive and multiperspective learning. Here all possible perspectives are considered along with all possible information to be used for effective learning. This actually helps in building whole-system intelligence to consider all aspects while learning.

To generate intelligence of, by, and for the whole system, all the parts of the system need to be aware of the decision scenario. In multiperspective learning, all different possible perspectives are tested with reference to decision scenario. The intelligence is built with the use of information coming from different perspectives.

In each phase of the cycle, there is a different relationship among different parts and parameters. *The relationships observed from different perspectives are preserved and knowledge base is built.* The perspectives and information from different perspectives are presented by different sources of information. In a typical case, these sources are different intelligent agents. *In the learning phase, this information is shared and weighted with reference to decision scenario.* This information includes information of events, knowledge of different parameters, different decision outcomes along with behaviors, patterns, and historical perspectives. These different perspectives need to be combined in order to build the overall decision context. Since the information becomes available over the time and even new perspectives are discovered—it is an ongoing process. This learning takes place in phases. The vision is information that can be directly derived from the data. Action is the information coming through exploration. The action and vision information comes from different intelligent agents or the sources of information. The iterative, collective, and cooperative learning take place to build the representative perspective based on this information. Whole-system learning also needs to be an iterative, cooperative, creative, collective learning cycle for MPML. Iterative means it keeps building knowledge with reference to the existing knowledge and revisits knowledge base in light of the new explored facts. It is collective and cooperative as it considers multiple perspectives and hence multiple agents need to cooperate so that all the parameters can be used and updated effectively.

Each one of the cooperative multiperspective learning phases contributes to build the overall context which can be represented as a multiperspective decision scenario. The perspective is associated with information flow and knowledge building. Information flows in and out of any given pool of information (database). Multiperspective learning is based on capturing this information and building learning parameters based on this information.

Multiperspective and whole-system learning can result using the learning strategies listed in following section.

The cooperative intelligence contributes by convening intelligent agents that capture different perspectives into a systemic decision scenario. But there are many challenges such as:

- The collective information building and representing the decision and learning scenario is a complex process.
- Generally, the decision process needs to be led by one of the intelligent agents, and it should have the capability or an algorithm that can give appropriate priority and weight to different perspectives which is a difficult task.
- The balance between exploitation of built knowledge and the exploration of new scenarios in a cooperative decision scenario becomes more complex as scope increases.
- The generalized models lack in considering the perspective and its effect to the required depth.

With reference to the above observation, we will try to explore the traditional techniques in the multiperspective light while system and cooperative learning use a different application. For any structure you need different perspectives to understand the complete three-dimensional view of the structure and without that learning is incomplete. Similarly for security applications, business decision making, image authentication, health diagnosis there are many perspectives and a complete learning system needs to consider all these perspectives and should make use of all parameters available in the system.

7.4   WHOLE-SYSTEM LEARNING AND MULTIPERSPECTIVE APPROACHES

As discussed above, it involves looking at the system holistically for learning. While learning, we are not looking at a particular part of the system. Furthermore, whole-system learning is using the parameters, experiences related to the whole system, and information available at various pockets of the system effectively. It is not just about use of the information but it is more about deciding learning policy and methods based on the whole system. It is not just one outcome value but a thought of multiple connections. It can be viewed as multiple sources of information coming together and optimizing for a particular decision scenario defined by the problem at hand. This considers all parts of the system, individual behavior of every part, and the behavior of them together as a unit. Whole-system learning takes place at different levels of system. The most important part of whole-system learning is a regular evaluation of decision parameters. This involves collaboration among information sources and collaborative learning among the multiple agents. Whole-system learning is about accessing more relevant information and using it more judiciously. The integration of information coming from the different sources leads to the whole-system picture. Generally, while learning, just some part of information and a specific part of the system plays a role. In whole-system learning every part of the system and every bit of information is used. This can even be described with a set of intelligent agents that capture information from different parts of the system and this information is used effectively for learning. We will discuss rough sets and a few other algorithms that can be used to solve these complex problems.

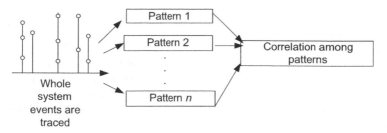

Figure 7.11   Whole-system event tracing.

WSML is about using and considering the information outside the direct experience or expertise. The perspectives and information made available by the multiple intelligent agents can help in making WSML possible. The whole-system parameters are used for learning. The whole-system events are traced and then the correlation among patterns can be used for learning. Figure 7.11 represents this aspect of WSML.

7.4.1   Integrating Fragmented Information

The major obstacle in whole-system learning is that the information comes in bits and pieces. The information is fragmented and hence makes it very difficult to build the picture of whole system and make effective use of complete information. Even in whole-system learning, one of the important aspects is dealing with fragmented information. There are efforts made to make sense of fragmented information.

7.4.2   Multiperspective and Whole-System Knowledge Representation

"Knowledge that is complex and ill-structured has many aspects that must be mastered and many varieties of uses that it must be put to. The common denominator in the majority of advanced learning failures that we have observed is oversimplification, and one serious kind of oversimplification is looking at a concept or phenomenon or case from just one perspective. In an ill-structured domain, that single perspective will miss important aspects of conceptual understanding, may actually mislead with regard to some of the fuller aspects of understanding, and will account for too little of the variability in the way knowledge must be applied to new cases. Instead, one must approach all elements of advanced learning and instruction, with the tenet of multiple representations at the center of consideration" [1].

The knowledge representation as discussed above can be done in the form of a decision matrix. This is the representative ID, but it is not from a particular decision perspective and hence all perspectives carry equal weight and the overall system information can be represented as a matrix.

7.4.3   What Are Multiperspective Scenarios?

Multiperspective learning takes into account multiple viewpoints. It encompasses the heterogeneous viewpoints, representations, activities, and roles within the system boundaries. These roles need to be considered in both collaborative and noncollaborative context.

The perspectives are a function of decision scenarios. For the business scenarios there can be perspectives such as optimization, profit, staff welfare, and so on. Similarly, for network and distributed systems the perspectives may include load sharing, security, flexibility, incremental growth, and scalability.

These perspectives in the case of decision making are also dependent upon the view of intelligent agent and information available with an intelligent agent. Different intelligent agents capture different perspectives and different parameters. These parameters and available information contribute to perspective of an intelligent agent.

7.4.4 Context in Particular

These multiple perspectives may not be in complete alignment with the decision context. Decision context in particular defines the scenario, the environment, and the objective of decision. The relevance of each perspective can help in prioritization of the information and building the representative decision matrix for the decision scenario. Figure 7.12 depicts multiperspective modeling for decision making.

7.4.4.1 Rough Sets: An Approach

- Multiperspective approach allows building of knowledge from more than one perspective. This will definitely improve the decision-making process. Most importantly, it allows using relevant information for decision making.

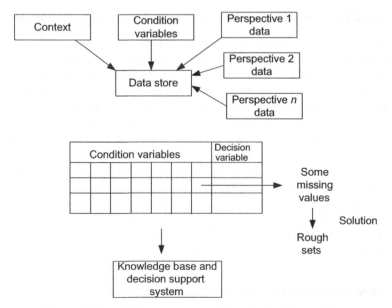

Figure 7.12 Multiperspective modeling for decision making.

- This increases the complexity; hence for the effective use of information for decision making, it is very important to consider decision context.
- The general perspective without context may not be relevant in a particular decision scenario, and hence it is most important to determine the right context and make use of it in decision making.
- Missing values and partial information can be used more effectively in decision context.
- Rough set theory can be used to determine missing values.

Thus a set is a collection of related things. Sometimes the nature of this relationship is not specified in these definitions. However, rough sets can be used in the case of imprecise data. We have discussed rough sets in detail in Appendix A.

7.5 CASE STUDY BASED ON MULTIPERSPECTIVE APPROACH

In this section, we will discuss some case studies on the multiperspective approach for better understanding.

7.5.1 Traffic Controller Based on Multiperspective Approach

Figure 7.13 depicts the multiperspective approach for traffic controller. Here task, laws, engineering, and issues are major parameters and different perspectives such as user perspective, legal perspective, and ecological perspective. All these perspectives with reference to these parameters build the complete system view.

The tasks include

- Traffic signal and overlap
- Ramp meter (freeway)
- Volume of traffic
- Congestion
- Pollution and fuel combustion

There are different traffic densities at different routes. The objectives from user perspective are minimum waiting time or no waiting time. Those from the ecological perspective are minimal pollution, and those from the legal perspective are smooth traffic and rules to be followed.

Layers:

Controller (strategic)→Field
Controller (supervisory)→Signal controller (local)

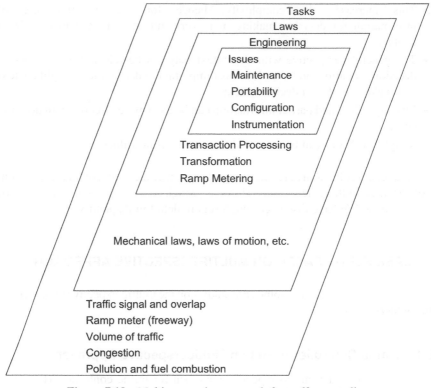

Figure 7.13   Multiperspective approach for traffic controller.

Laws:

    Mechanical laws
    Embedded Kirchhoff's laws in current electricity
    Intersection and parallel law in current electricity

Logics:

    Ladder logic (electrical model)
    Transaction processing (state machine)
    Transformation (embedding laws or simulations)
    Procedural programming (H/W or assembly)
    Ramp metering (block removal)

All these parameters build the overall parameter set. With reference to a particular perspective, a few parameters are important while others are not that important. The effective use of these parameters in a decision scenario can be used for whole-system multiperspective learning.

7.5.2 Multiperspective Approach Model for Emotion Detection

By considering emotions as a means of communication, human computer interface (HCI) shall become more natural—that is, more effective in human interactions, where information is transmitted not only by the semantic content of words but also by emotional signaling in facial expression and gesture, which forms the basic single perspective approach. Thus in human–computer interaction systems, emotion-recognition systems could provide users with improved services by being adaptive to their emotions.

In real-life scenarios, perspective and in effect the requirements of data are changing dynamically. Knowing the real relevance of data with a particular context is not an easy task, as we have to take into account users' wishes or needs, as well as the underlying physical and social context. The purpose of quantifying the performance of single-perspective systems is to recognize the strengths and weaknesses of these approaches and to compare the different approaches that can fuse these dissimilar modalities to increase the overall recognition rate of the system.

If we judge the emotions of a person only on the basis of either his expressions or his gestures, though it gives us some idea about the corresponding emotion, it is not always an accurate one. However, by viewing the subject of emotion detection from various perspectives and taking the cumulative result into consideration, it is bound to give us a more precise and exact judgment of the emotion.

7.5.2.1 Understanding the Context
In most exercises of recognition, it is helpful to have cues that might lead to accuracy in recognition. These cues are provided to us by the emotional context of the person. For instance, when a person is crying, all the detection technique will tell you is that the person is crying. But it is difficult for the machine to determine if the person is crying because of happiness or is he genuinely sad. So, this is where the context of the situation comes into picture. The context of the person could refer to the environmental context, the emotional context, the social context, or even the emotional context (perspective) of the person under examination.

In terms of context, it is worth mentioning that the mood of a person forms the underlying context, considering which we can judge their present emotion or reaction. It is this very context which must be observed and taken into account to enhance the quality of understanding.

In a study by Righart and de Gelder [2], participants were presented with images of faces depicting happy, fearful, and neutral expressions, paired with varying background environments that fit one of the same three categories. This study supported the idea that environmental context and visual cues can enhance the recognition of facial expressions, as facial expressions were more quickly recognized when in a congruent emotional context a background image that matched, in tone, the emotional expression was present in the face.

The effect of social-scene context was also observed. Bodily expressions were better recognized when the actions in the scenes expressed an emotion congruent with the bodily expression of the target figure. The specific influence of facial expressions in the scene was dependent on the emotional expression but did not necessarily increase the congruency effect. Taken together, the results show that the social context influences our recognition of a person's bodily expression.

7.5.2.2 Different Approaches Toward Emotion Detection

Emotions have a profound effect on various facets of human nature. As a consequence, there are a lot of parameters that can be taken into consideration for judging what the person is feeling and to what extent. These attributes are: facial expressions, gestures, body signals, and speech.

Using Speech (Audio Source) The rationale behind emotion studies in connection with human–computer interaction is to build a machine that serves users' needs in a more natural and effective way. The work done on detecting emotion in speech is quite limited. Currently, researchers are still debating what features influence the recognition of emotion in speech.

One of the main challenges faced during emotion detection using speech is that the system has to work independent of the speaker. So those features of speech are still being researched which encode maximum information about the emotional state and are just a result of variations in human voices.

Emotions influence a lot of parameters reflected in human speech. Bäzinger argued that statistics related to pitch convey considerable information about emotional status [3]. Pitch values, however, exhibit a large amount of variation between speakers. Hence, input parameters such as pitch have to be normalized before they can be used. This normalization is done by building a cumulative histogram of each property, constructing a normal distribution, and mapping each individual voice into this distribution.

Another parameter in use is MFCC. In sound processing, the *Mel-frequency cepstrum (MFC)* is a representation of the short-term power spectrum of a sound. So *Mel-frequency cepstrum coefficients (MFCC)* are used as features of the speech but are independent of the speaker and its gender. MFCCs are the most widely used spectral representations of speech. Kim et al. argued that statistics relating to MFCCs also carry emotional information [4].

The probable system consists of four major steps:

1. Speech acquisition
2. Feature extraction at each timescale level
3. Machine learning for each feature set includes the use of K-means clustering
4. Information fusion to merge the information

However, when considering speech as a basis for emotion detection, there are many exceptions that need to be considered to avoid incorrect interpretations. For example, the word "scheduled" may be pronounced in different ways by a person of US origin and of British origin. Thus, the accent of a person greatly influences the phonetic interpretation of speech. On a related note, phonetic information that is determined from the emphasis or sarcasm in speech is not necessarily consistent or precise.

Needless to say, audio sources are generally susceptible to interference (i.e., noise) from the external environment. Furthermore, biological variations due to nature of voice of each individual are difficult to clearly state or monitor. Also, timely variations cause deviation from the norm for even the same individual at different times. For instance, a person suffering from cold or flu may record different modulations before, during, and after his/her illness. To comprehend his/her emotional state based on speech alone would lead to incomplete or inconsistent results on being compared with the other perspectives.

Using Facial Expressions (Visual Source of Input)   Facial expressions are a powerful tool of communication amongst all living beings, particularly us, humans. Considering this, perspective is a vital part of gathering information about what a person is trying to communicate to us.

There is a unique expression made for each unique emotion felt. There are over 20 different facial muscles. The basic essence of a facial expression is that when contorted in different ways and combinations, our facial muscles perform the seemingly simple task of expressing happiness, sadness, anger, and several other feelings.

While studying facial expressions, we look at some particular areas of interest— that is, the eyebrows, the eyes, the mouth region, the nose, and the chin. When breaking down an expression, we can look at each *region of interest* and specify the change expected there in order to depict or recognize a particular emotion. Emotions or expressions are described to a computer system in the form of deviations from a neutral baseline for each of these regions.

An obvious parameter to consider is geographical variations of facial features such as eyes, mouth, nose, etc. For instance, people of East Asian origin have predominantly narrow eyes, small noses, and other discernable variations in features. No matter how general our approach to identifying facial expressions may be, some level of personalization is necessary in order for greater accuracy in the results. Furthermore, accidents causing distortion of features or even paralysis will affect the data input to the system and hence lead to inconsistent results. This may even be caused due to drug medications or drug abuse in the subject under observation.

It goes without saying that facial expressions play an important role in communication. Not only can visual information confirm or contradict the information gained from other perspectives, but visual knowledge is also sufficient for reaching a reasonable conclusion about what is perceived. Moreover, as far as emotions are concerned, very rarely does a facial expression contradict what the person is actually feeling. When it does, the subject concerned is usually lying and very adept at doing so, which necessitates consulting the other perspectives for a more wholesome judgment.

Using Hand Gestures   Hand gestures form a very expressive part of peoples' interactions. Changes in hand gestures and body language can say more about the ongoing conversation or interaction than the actual words. For example, normally a number of people play with their hands or tap their feet when they are impatient or nervous. But this approach of using hand gestures has issues such as regional and cultural traditions and involuntary personal habits. Thus a thorough analysis is required to come to a more accurate result.

The main problem in this approach is the wrong interpretation of some gestures. For example, a person may be analyzed as nervous due to the position of his arm as per norms, but the case might be that he is habitual in doing so due to social and cultural factors. Once again, biological distortions play an important role in hampering the process of emotion detection and recognition. Thus a greater margin for wrong inference of gestures is possible due to context-based inadequacies.

Using Biosignals It's a natural phenomenon that an emotion such as joy is systematically associated with increased body temperature, heart-rate acceleration, sensation of lump in the throat, and some muscle symptoms. Similarly, anger is marked by sharper increases in body temperature, heart rate, perspiration, breathing rate, and muscle tension. Also, sadness is associated with less bodily sensations, mainly in the stomach and throat areas. Thus, such changes can be recorded and used to recognize the relevant emotions, as per the data already available.

A number of monitoring mechanisms exist in hospitals (such as ECG, EEG, EMG, BVP, etc.) which tell us about the reactions the body makes to support the emotion felt. These mechanisms are employed to better understand how the body reacts to the different emotions felt. The polygraph test deserves mention here, since it also employs the above mechanisms to detect lies and pretense.

However, this information alone is not sufficient to entirely judge a person's state of mind, since particular signals could mean a number of things. For example, raised heart rate could imply excitement, but could also indicate nervousness or anger or even fear. Also, the reactions observed in a body are instantaneous and are very volatile in nature which is in tandem with the fickle nature of the human mind. But, this brings about a lot of hardware constraints; there is a need to constantly monitor changes in the human body.

This approach, however, may prove to be inaccurate under various scenarios and context, for example, change in body temperature due to external factors like climate and state of health.

7.5.2.3 *Model for Emotion Detection* Figure 7.14 depicts a multiperspective model for emotion detection. Here different perspectives are captured for detecting emotions. In the above multimodal model approach, facial expression, speech analysis along with hand gestures, and body language along with temperature are used together to get individual results and conclusions. At the decision level, a comparator and an integrator structure can be used which handles the results from the various single-modal approaches. This helps in removing inconsistencies and eliminating inaccurate interpretations. It also helps in coming to a decision instead of missing values or conclusion from the single-modal phases. The problem of halting

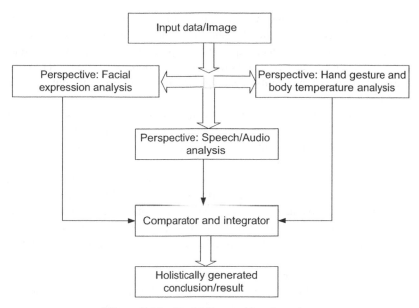

Figure 7.14   Model for emotion detection.

or false-result generation due to missing values of the modal phases is eliminated by the outputs of the other modal phases. For example, due to various reasons the gesture analysis has some missing value of input and is unable to generate a valid output. This can be neglected because we have other means of getting to the conclusion due to the other single-modal approaches that are fairly independent from the hand-gestures analysis one.

The holistically generated conclusion takes into consideration the output of the integrator as well as the context of the input, which can be the background as well as the previous analysis and result. Thus a sound and fairly accurate conclusion can be reached, increasing the efficiency of the emotion-detection system.

7.5.2.4 *Integrating Perspectives and Generating a Holistic Output*   For deducing the final emotion, we will try and solve the problem from different perspectives, as discussed until now. The next step is to collectively look at the different perspectives and integrate them to yield a final conclusion about the person's felt emotion. A possible way to do this is to use the branch-and-bound technique.

Understanding these perspectives shows that some perspectives tend to be more reliable than others, and we use this knowledge to prioritize them. More precisely, we can assign different weights to each perspective and consider their output in decreasing order of their reliability. For example, if we conclude that facial expressions are the most reliable way of understanding human emotions, we may consider what they tell us before looking at the audio samples or biosignals. Hence, we can apply facial expression analysis in stage 1, speech analysis in stage 2, followed by gesture-analysis in stage 3, and so on. If stage 1 tells us that the person is crying, we can use stage 2 to decide why (due to genuine sadness or extreme joy) he is crying. Within the branch-and-bound technique, we can perform this by taking one stage at each level of the state space tree. In this way, each leaf node of the tree will represent a possible combination of the results of each perspective, applying different priorities in each scenario.

While considering more than one perspective in judging emotions, it is certainly plausible that different perspectives may lead to different opinions about the emotion being felt. In this case, stage 2 may contradict the analysis that stage 1 provides us with. When this happens, the system will have to decide depending on each perspective's degree of reliability and assign weights to the intensely emotion felt.

Here, the concept of mixed emotions comes into picture. Many times, people feel mixed emotions, and thus the signals received from them may pose a rather confusing picture to the system trying to interpret them. Considering different perspectives will help in deciding which emotion is felt to a larger extent, along with some suppressed emotions.

The final output can be related in terms of the mixed emotions, depicting how strongly each may be felt in the person being observed. For example, a person may be excited as well as happy, and the result may be in the following form: 60% excitement, 40% happiness.

7.6   LIMITATIONS TO A MULTIPERSPECTIVE APPROACH

As discussed, considering a problem from different perspectives broadens our understanding of its nature. However, in spite of it being a very effective approach, it poses great difficulty in actual implementation due to the reasons stated below.

In order to implement a multiperspective model for a problem, each perspective must be accurately described and implemented in itself first. This requires more the processing power, since processing must be done not only for each perspective, but also for integrating and combining their approaches towards a unanimous result. Naturally, the time taken to apply this is several times greater when compared to a single perspective. The resources and hardware required for this are extensive and expensive. Such heavy-duty implementations will not be suitable for simple applications such as room ambience even though they may be very valuable for something such as criminal psychoanalysis.

7.7   SUMMARY

Multiperspective learning and decision making is one of the most important aspects of systemic machine learning. The knowledge about single perspective or information from one perspective may not result in effective learning. Each perspective offers some additional information and more insights into system dependencies and relationships among subsystems. The information is always fragmented, and there is a need to build a complete picture of the system for decision making based on this fragmented information. Use of PDSRD can allow representing a particular perspective of the system. Whole-system learning is making the use of all information available to make a decision. Furthermore, this information can be used to represent the overall system decision dependencies.

Whole-system learning combined with multiperspective learning allows making use of all the available information with reference to a particular decision scenario. These perspectives and representations build the knowledge for systemic decision making. The systemic knowledge can help in determining dependencies among features, actions, and impacts. The knowledge built can even be updated with information that becomes available from a new perspective. Multiperspective and WSML allows building the overall system knowledge for the effective use of available information.

There are many perspectives possible even for a simple problem, and hence the number of dimensions of learning problem keeps increasing. These very high dimensions can complicate decision scenarios without adding benefits in some cases. Hence there is a need to prioritize parameters and perspectives with reference to the decision scenario. Another challenge is selecting the relevant information and combining this information for learning. In short, different statistical methods can be used to prioritize perspective. Multiperspective and whole-system learning aims to present all the available attributes, with their selection and prioritization. Furthermore, it attempts combining perspectives with reference to decision scenarios. With the complete picture at hand and all information and historical patterns available, efficient learning to deal with complex scenarios is possible.

REFERENCES

1. Jehng J and Spiro R. Cognitive flexibility and hypertext: Theory and technology for the non-linear and multi-dimensional traversal of complex subject matter. In D. Nix and R. J. Spiro (Eds.), *Cognition, Education, and Multimedia; Exploration in High Technology*, Hillsdale, NJ: Lawrence Erlbaum, 1990.

2. Righart R and Gelder B. Rapid influence of emotional scenes on encoding of facial expressions: An ERP study. *Social Cognitive and Affective Neuroscience*, 2008, **3**(3), 270–278.

3. Banziger T and Scherer K. The role of intonation in emotional expression. *Speech Communication*, 2005, **46**, 252–267.

4. Kim S, Georgiou S, Lee S, and Narayanan S. Real-time emotion detection system using speech: Multi-modal fusion of different timescale features. *Proceedings of IEEE Multimedia Signal Processing Workshop, China*, 2007.

CHAPTER 8

Incremental Learning and Knowledge Representation

8.1 INTRODUCTION

In case of supervised learning, the representative training data plays a key role in deciding the performance of learning algorithm. This representative training data may or may not represent what exactly it is expected to represent. Furthermore, in due course, more and more data, and more and more information becomes available. This new data may bring new perspective, may even change the statistical distribution of the data, and may even compel to revisit the known premise. Understanding the importance of new data and allowing it to play a suitable role in learning to improve a task is tricky. Retraining the learner with all data and discarding all old learning is one of the approaches followed in this case. This approach has many limitations in terms of efficiency and knowledge retention.

Human beings use already possessed knowledge along with the experiences for learning and decision making. When a person comes across the new event or information, he learns incrementally without discarding existing knowledge. This incremental learning (IL) tries to verify the existing hypotheses and also in this process formulates a new hypothesis. Knowledge is accumulated incrementally and represented in such a way that incremental learning becomes possible. Practically speaking, incremental learning is one of the major strengths of human beings. The learning begins initially based on the facts available; as the new facts become available, the overall knowledge is refined. Rarely in this new scenario is the complete reformation of knowledge required in case of human beings. Learning is about using the existing knowledge and the new available information and building the most efficient knowledge base. Another most important aspect of learning is the representation of knowledge. This representation should allow accommodating or using the new information effectively for learning. Incremental learning should allow not only accumulation but also updating of knowledge in light of new facts revealed; and while doing that, it is not supposed to lose the useful knowledge built in the past.

Reinforcement and Systemic Machine Learning for Decision Making, First Edition. Parag Kulkarni.
© 2012 by the Institute of Electrical and Electronics Engineers, Inc.
Published 2012 by John Wiley & Sons, Inc.

Some of the obvious reasons for incremental learning in the case of humans are the memory limitations and the sequential nature of the received information. Still this is the most efficient way of learning known so far. In all real-time complex systems, we need the efficient method and incremental learning ability to cope up with knowledge-retention challenges.

In this chapter, we will discuss incremental machine learning and knowledge representation. Can we make machine learn incrementally? This is the key question that we will try to answer in this chapter. There are different learning approaches to make decisions. Generally, most of the machine-learning approaches lack the ability of using the knowledge which is already available during the next phase of learning. This is one of the most important factors, and not having the ability to learn incrementally may cost us in terms of efficiency as well as knowledge. "Learning every time from the scratch" leads to serious limitations on learning abilities of the system. These limitations are mainly observed in terms of abilities to hold knowledge and efficiency to handle learning complexity. With more and more information available at our disposal and every day more information becoming available, it is highly likely that intelligent systems would like to make the best use of all the information available. During the initial training, any system is trained on the known facts and available training sets. As the time progresses through exploration and through other information sources, more and more information becomes available. This information may be in alignment with previous premises or sometimes force us to change premises. Incremental learning is about making the best use of available information without losing the useful knowledge acquired in past learning cycles and, while doing this, correcting the premise if proven wrong.

Incremental learning is not just the ability to keep learning from the new data whenever it becomes available, but it is also about testing the hypothesis based on new learning. Each approach has its own identity and is used depending on the type of application and the nature of outcome required. Considering the growth rate of the data, new methods with great potential for decision making in terms of classification and accuracy are required; at the same time the response also needs to be fast. With this, there is a need for incremental learning, which will add value to the existing learning methods, and where the learning takes place on top of already learned data.

One approach used for incremental learning is generating an ensemble of learning techniques and classifiers based on the dataset when it becomes available. These ensembles of classifiers are combined using weighting and voting or some other similar mechanisms to get the best out of them. The weighting can be static or dynamic. In real-life scenarios, dynamic weighting makes more sense. In this chapter, we will discuss the various incremental learning methodologies and the need of incremental learning from the systemic machine-learning perspective.

8.2 WHY INCREMENTAL LEARNING?

With the supervised learning method, the dependency of the classifier lies on the training data available, whereas in case of unsupervised methods the classification is done with the unlabeled data. Unsupervised machine learning is based on similarity and closeness and definitely exhibits a few incremental learning properties. The fact that datasets may be evolved at a later stage remains unfocused. The datasets, relationships, and even the parameters are evolved over the time. In the absence of the knowledge of the new scenarios and if the new relationships are not explored among the data, the intelligence and decision abilities will be restricted by the initial training sets.

The data could be training data or unlabeled data which is generated over a period of time. Consideration of these data will have an impact on the decisions made earlier and may allow the improvement in overall mapping. Furthermore, not only the relationships among data points but the relevance of that data with reference to learning scenario is important. The other factor that needs to be looked upon is the time spent in training. In the case of huge training sets, this time is considerably high. A new learning method that would be swift and efficient is required. The learning that is based on complete data and does not use an incremental approach may simplify the situation in some cases—but in most of the practical scenarios, it not only takes more time but also limits the learning ability. Imagine a person who asks you the complete story from the beginning in case of any small new revelation. This will not only irritate the listener but also limit his learning abilities to a great extent.

Another aspect is that at each stage of learning, some knowledge is generated. This knowledge may be some relationship, pattern, or even correlation. The effective use of knowledge built in previous stages remains unnoticed in the absence of incremental learning. Exploitation and updation of this knowledge for better decision making is a critical factor. So the feature vectors that are generated need to be updated with new ones. The knowledge built in each learning cycle is important, and new learning strategy needs to use the relevant knowledge from those learning cycles in light of the new information that becomes available. The hypotheses used at every stage should not contradict the data at any stage. The intermediate hypotheses and learning need to be maintained as they can contribute to the knowledge base.

Considering the above-mentioned limitations, it is clear that the learning takes place in phases, and during every phase the learning algorithm gets some new data and new material for learning, which creates the need of incremental learning. The incremental learning is required to offer swift and accurate decisions. *IL is about effective use of already formed feature vectors or knowledge base during the next phase of learning without impacting on accuracy of decision making.* Figure 8.1 depicts the factors that put forth the need for incremental learning. Here incremental learning has various important parts such as knowledge updating and reuse during every learning cycle. Decision making allows the most important exploration where feedback is used for incremental learning. Identifying the new datasets that are useful for learning and then learning from these datasets is the most important part of the process. Incremental learning also needs knowledge updating, incremental decision making, and learning based on time, efficiency, and accuracy tracking.

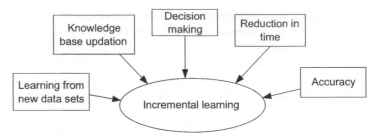

Figure 8.1   Factors accelerating the need for incremental learning.

8.3   LEARNING FROM WHAT IS ALREADY LEARNED. . .

What the system has already learned may make more sense in case of new revelations or on availability of new data or information. In some cases, in light of the new data, the hypothesis based on previous dataset may not make sense. Incremental learning would rightly be described as the *ability of the learning methodology to make effective use of new information* and already formed feature vectors or existing knowledge base which is generated in a previous phase of learning. Unlike the various methods available for classification, the intention of incremental learning is to exploit as much knowledge as you can and get fast and accurate classification. Figure 8.2 depicts this scenario.

Generally, what we have learned is used in decision making quite efficiently. But interestingly it is not used for learning. The very reason for this is that we are keen in putting the complete data together for learning. Some methods stick to the old hypothesis, while the rest of them try to come up with a new one. This results in relearning the same facts again and again or sometimes ignoring some important knowledge that is already learnt. The idea of incremental learning is using effectively what is already learned. Can incremental learning be absolute? Absolute incremental

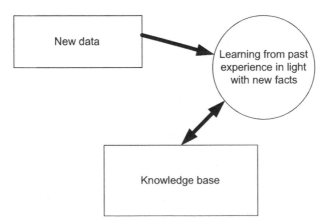

Figure 8.2   Learning fromwhat is already learned.

learning comes with its own limitations and does not give the opportunity to correct the premise during learning. Incremental learning can be broadly classified under two categories—absolute and selective. Here absolute incremental learning does not look into old premises, nor does it verify or correct what is already learned. Selective incremental learning, however, takes care of these issues.

8.3.1 Absolute Incremental Learning

Absolute incremental learning can be referred to as a traditional approach to incremental learning. In this type of approach, the new data are analyzed separately; new feature vectors are formed and are combined with the existing ones. Here the knowledge built by the classifier is called as knowledge base. Thus the knowledge is updated and is used further for classification. Figure 8.3 shows the basic view of absolute incremental learning.

Though this approach updates the knowledge in an incremental way and is a very efficient one, it has some limitations:

- Knowledge about which feature vectors to use is not available. It simply adds up to the knowledge base. This learning builds too much of redundant information and cannot even maintain relationships among different outcomes.
- Impact of the new data on the already formed vectors is difficult to determine. Hence it may fail in handling boundary conditions.
- Sometimes addition of the data to the vectors results in complexity unnecessarily. This results in a huge number of feature vectors increasing the complexity of learning and decision making multifold.
- There might be a case where some feature vectors are not required further or are invalid. Discarding these vectors becomes very complicated. The impact of discarding these feature vectors can even be felt on other parts of the system.
- With an increase in the size of the knowledge base and more and more feature vector formulation, it leads to ambiguous states. Handling boundary conditions becomes rather complex.
- There is a need for use of semisupervised learning effectively for better outcome, which does not persist with absolute incremental learning.

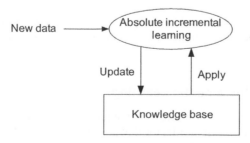

Figure 8.3 Basic view of absolute incremental learning.

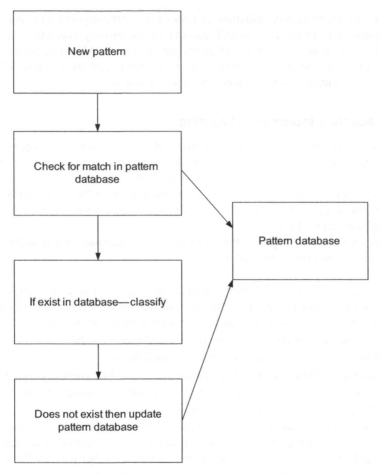

Figure 8.4 Absolute incremental learning.

Absolute incremental learning is useful in the case of simple learning scenarios where the extent of interdependencies among feature vectors are very low. This is especially useful in cases where addition of new feature vectors does not have any impact on knowledge base built in the past. Figure 8.4 depicts absolute incremental learning.

Though absolute incremental learning is simple and efficient, it has many limitations. The limitations of absolute incremental learning pose the need for selective incremental learning.

8.3.2 Selective Incremental Learning

To overcome the pitfalls of absolute incremental learning, the learning now needs to be selective in nature. The major factor that absolute method falls short of is a strong decision support engine and the ability to distinguish between incremental learning and nonincremental learning scenarios. In all the cases, the system need not be incremental while learning. It should retain the useful knowledge while updating the rest of the feature vectors in light of new revealed facts. Furthermore, it should refine a few already learned feature vectors in light of the new knowledge and may keep rest of the feature vectors unaltered. Selective incremental learning refers to learning incrementally in a selective fashion, selective areas, and selective scenarios. It is adaptive responding to the current changes in the system and at the same time maintains the accuracy at acceptable levels. The selective incremental learning flow is depicted in Figure 8.5 with reference to pattern-based learning. When the new pattern is observed by an agent, the pattern is checked for the similarity with reference to patterns in pattern database. In the case where a similar behavioral pattern is observed, the new

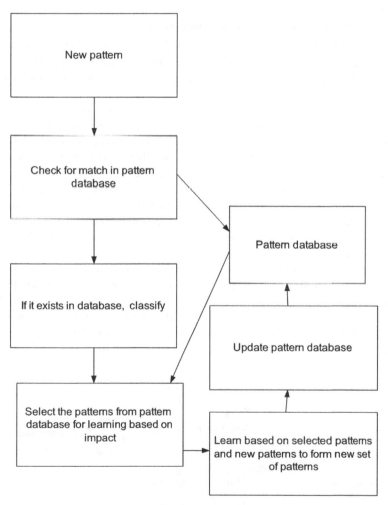

Figure 8.5   Selective incremental learning.

pattern is classified to the respective class. But in the case where there is no pattern or no class to which it could be classified, a set of patterns from pattern database and corresponding data from training set are selected—based on the impact of including the new pattern. The system is trained for the new pattern with reference to selected patterns and training set.

The properties of selective incremental learning are

- Learn from the new instance that is evolved. It neither learns from scratch nor keeps all feature vectors learned previously as they are It evolves the existing feature vectors to accommodate new datasets.
- Learn from the new training set that is evolved during the course of time.
- Update the feature vectors selectively with reference to the impacts of new scenario.
- Discard the scenarios that are no longer valid.
- Formulate the feature vectors incrementally for decision-making scenario.
- Be selective in nature; that is, it will select between incremental and nonincremental learning which is based on the analysis of the new data available.

Learning is not a mechanical activity involving collection of feature vectors and applying some standard learning algorithm for learning and mapping of these feature vectors. It is more selective and dynamic in nature. Creation of initial body of knowledge and refining it selectively and dynamically in light of new knowledge and learning scenarios is selective incremental learning. Figure 8.6 shows selective incremental learning with decision support engine that makes it act selectively for decision making. As said earlier, selective incremental learning has to be selective and adaptive in nature. It is adaptive to the current scenario or current system state. This adaptive nature helps in having a better decision-making capacity for forecasting methods as well.

In the case of selective incremental learning, the selection considers many inputs such as which data should be considered for incremental learning, whether the learning needs to be incremental in a specific scenario, and the selection of method. We will also refer to this learning as Dynamic Selective Incremental Learning (DSIL). A typical DSIL is depicted in Figure 8.6. In this scenario, the initial learning is confined by the training sets. The knowledge base is built in the process of learning

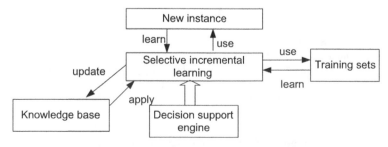

Figure 8.6   Basic view of selective incremental learning.

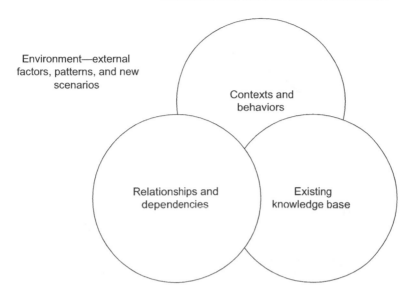

Environment—external
factors, patterns, and new
scenarios

Contexts and
behaviors

Relationships and
dependencies

Existing
knowledge base

Figure 8.7   Incremental learning factors.

and the decision support mechanism with reference to knowledge base delivers decisions. In the case of a new instance, the decision about learning is taken selectively. Here the impact of new scenario and learning with reference to the knowledge base is identified, and only the selective part of the knowledge base is updated to accommodate the new scenario.

In the case of incremental learning, it is expected that knowledge building should take place. Hence DSIL should consider different perspectives while learning. Figure 8.7 depicts selective incremental learning and the consideration of different perspectives and contexts while learning. Generally, while selecting a learning algorithm, the best one fitting to the need is selected. Hence obviously more than one learning algorithm may improve the overall accuracy of the learning and classification. Each of the learning algorithms has its own assumptions. If these assumptions do not hold for the data at hand, then it may lead to some sort of error. The learners are fine-tuned for the given set of data.

Due to the complexity and failure to consider some of the perspectives, even the best learners fail to produce accurate results. A simple selective incremental learning algorithm tries to identify the region or the part of training set which may be impacted due to realization of new facts.

Figure 8.8 explains the selective learning process with reference to actions that results in new data. These new data can be of different types:

- Data that are very similar to the data in training set and does not need any new learning to handle it.
- Data that are completely new and which have never been seen by the learner. But these new data produce a completely new pattern or rather create a complete new class and follow a new decision-making process.

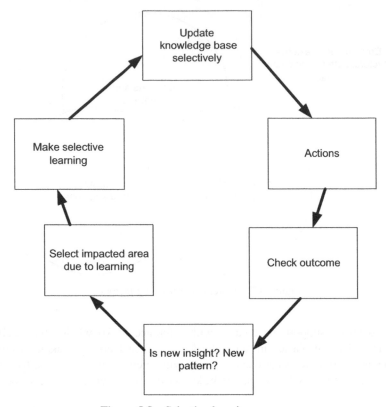

Figure 8.8   Selective learning process.

- Data that are similar to the data from the training set but demand different course of action and in that case the decision boundaries of a particular training cluster need to be redefined.
- Data that have the impact on the complete learning policy and reveals new facts about the system.

Collective incremental learning/cooperative learning is learning between different parts of the system and intelligent entities incrementally. It can typically be learning between teams and different learning and information parts of the system. Since selective incremental learning needs the information about dependencies of data points, there is the need of learning elements to interact. Hence to make selective learning possible, there is a need for collective learning. The concepts perceived by an individual agent need to be objectified. Collective incremental learning is concrete learning. Selective, collective, automatic learning leads to collective incremental learning. It is depicted in Figure 8.9. The collective incremental learning allows us to determine the region for selective incremental learning with reference to different intelligent agents. Hence the collective incremental learning needs to be collective,

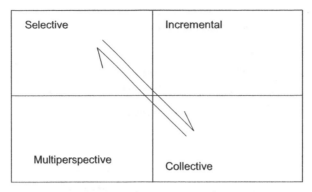

Figure 8.9 Collective incremental learning.

selective, incremental, and multiperspective. The new information or new data point has impact on certain part of the system and this region is selected collectively with multiperspective analysis. Then based on this selected region incremental learning is performed.

To achieve better accuracy, incremental learning takes place at different levels. It can be region specific, it can be between two agents, or it can even be across the system.

Different conditions, rules, and interdependencies are the levels of complexity indicating levels of learning. Incremental learning generally takes place in case of a new event or data point. These learning events or explored dependencies are analyzed in the case of incremental learning.

The incremental learning can take place at different levels. Figure 8.10 depicts the same. These levels are controlled by the dependencies. The conditional probability tells about these dependencies and impacts. Figure 8.11 depicts these relationships with reference to events. The parameters and information that become available through the events can be used for learning. The instructional event is about

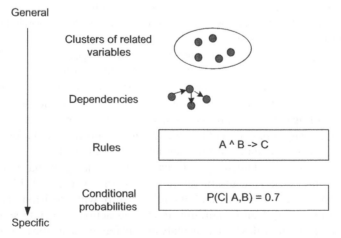

Figure 8.10 Learning at different levels.

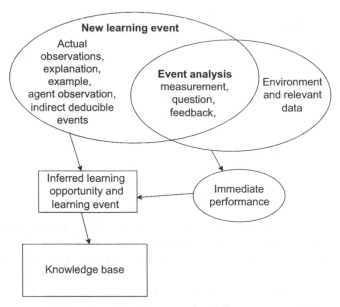

Figure 8.11   Learning events.

explanation and practical observations providing the information about the event, while assessment events are events that provides feedback.

Knowledge refinement with reference to incremental learning is represented in Figure 8.12. Through new information and knowledge parameters, a sense is made and that is used to take a decision about knowledge retention, delearning, and relearning. There are new data points, new observations, changes in behavior, or visible changes that typically contribute to the learning event. There are direct learning events and indirect learning events. The learning opportunities also result from the analysis of events or measurement or interpretation and inference. The learning opportunities and parameters are used with reference to knowledge base for collective incremental learning.

The knowledge is built, and this contribution to knowledge comes through learning based on experience and use of new unlabeled data. Here learning event is any event that helps to build new knowledge, it is the action or decision that can provide ground for learning, This knowledge is used in terms of outcomes and decisions. This knowledge building continues through feedback from the users and outcomes during different phases. Any application of knowledge creates an opportunity for learning in the form of a learning event, and this can help to build knowledge. In a collective scenario, knowledge is retained by a different intelligent agent that is transferred based on learning needs and ultimately helps in learning. Figure 8.13 depicts the process of knowledge building and its relationship with decision making and inference. In the case of an exploration mode, the actions generate some sort of feedback. Observations, rules based on analysis, and feedback are the contributors for learning. This takes place in association with the system or environment.

Knowledge transformation

Learning processes			
Applying knowledge	Learning element with systemic inputs facilitate knowledge application	Learning element with systemic inputs can be adapted to new needs	Learning element with systemic inputs can derive more knowledge
Refinement	Knowledge retrieval	General inference apply across system	Knowledge building block
Knowledge building	Prioritization— high-priority learning events	High-priority learning event— their analysis and cognitive adaptation	High-priority learning event based learning

Retention → → → **Learning**

⇩ Knowledge retention   ⇩ Knowledge transfer   ⇩ Learning

Figure 8.12   Incremental learning and knowledge refinement.

Incremental learning is possible through the use of knowledge built with reference to new events and identification of knowledge-building opportunities. Any new event is classified using the knowledge base and event histories. Figure 8.14 depicts event generation and knowledge consumption and with reference to them how incremental learning and knowledge building takes place.

Let us consider an example of classification of documents to have a better understanding of absolute and selective incremental learning. We have two sets of document classes, say history and politics. A new document is to be classified. With absolute incremental learning, the knowledge base is updated with the feature vectors

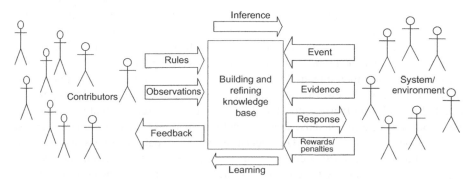

Figure 8.13   Inference and knowledge building.

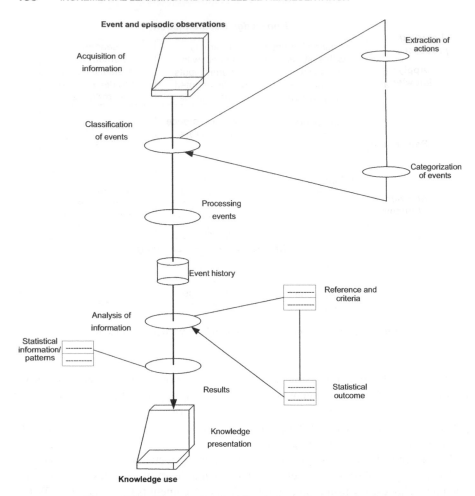

Figure 8.14   Process of embedding information and knowledge.

and the classification would take place in an absolute way between the given set of documents. In this case, the learning will not allow updating of existing feature vectors, in case of failing to classify the new set; a completely new feature vector and class will be formed for that information. It fails to consider the impact of this new class on the existing set. Consider the new document set has word "politically" in it. With this word, the absolute incremental learning would classify it in politics and update the knowledge. But with selective incremental learning, it learns with reference to the selected part of data that is impacted due to the new dataset and feature vectors. It would possibly be a case that the document actually belongs to history or there is a need for creation of a new class. This decision-making capability exists in selective incremental learning. Further collective incremental learning can help in understanding the document class with reference to interactions, with relationships among all classes removing the limitations of absolute incremental learning.

One thing to always remember is that the incremental learning is a continuous process. It is on all the time; whether new data are available or new classes/labels are generated, the learning is always "*active.*" DSIL keeps track of changes in relationships among feature vectors with revelation of the new information and new knowledge.

8.4 SUPERVISED INCREMENTAL LEARNING

Incremental learning is about responding to new information effectively without retraining. After the discussion of incremental learning, let us move to supervised incremental learning. In a typical supervised learning methodology, we have labeled data for our reference that is used for learning. The learning results on the basis of training using the labeled data. We train the classifier on the basis of this labeled data. Further new data are classified on the knowledge built on the basis of labeled data used for training. In the case of supervised incremental learning, the training set needs to be enhanced without retraining the complete training set. In short, new data can be used for learning and that is done through enhancement of learn sets.

With incremental learning, the learning will take place in a semisupervised way with the existing supervised learning method. There is a subtle difference between semisupervised learning and incremental supervised learning. In supervised incremental learning when additional training set is introduced for learning, it is incrementally included in the complete training set. There are two approaches in supervised incremental learning as discussed above. In absolute incremental learning, the existing feature vectors are not refined and the whole learning approach is incremental. It is very useful in case of limited boundary cases. In selective incremental learning, the selected training set based on the proximity and impact of new data and new information are retrained in light of the new information.

The incremental supervised approach performs the following tasks:

1. With the training data, build a knowledge base. (Any supervised algorithm does this.) This is also done in such a way that the representation of knowledge helps in making quick decisions ahead.
2. With the unlabeled data, classify and update the knowledge base.
3. For data leading to ambiguous classes, select an optimal solution.
4. With new training data available, make a decision to update/restructure the knowledge base.
5. If required, generate new classes, or merge existing ones and reorganize the knowledge base.

8.5 INCREMENTAL UNSUPERVISED LEARNING AND INCREMENTAL CLUSTERING

In unsupervised learning, the learning results are based on similarity and differences, closeness, and distances. Clustering is simply grouping of unlabeled data based on similarity. We are aware of the clustering methods—hierarchical and nonhierarchical (k-means). The time factor in the case of clustering is a point to be looked upon. With a lot of unlabeled data, the clustering needs to be swift and should maintain the accuracy of the clusters formed. In hierarchical clustering, multiple steps are used for clustering and the data are not partitioned into a particular cluster in a single step. In this case, a series of partitioning is used. It may begin with a single cluster with all objects slowly divided into a number of relevant clusters. Hierarchical clustering has a set of agglomerative methods, which proceed by a series of agglomerations of the n objects into meaningful groups, and divisive methods, which separate n objects successively into finer relevant groupings. Hierarchical methods are generally sensitive to outliers. The number of clusters selected should be optimal. With k-means, the requirement of the number of clusters in advance plays a crucial role in the final outcome. The movement of the data from cluster to cluster until a stable state is reached is time consuming. Figure 8.15 depicts a typical cluster formation and distance measurement in case of clearly separable clusters.

In the case of partition-based clustering, initial partitioning of data points or objects is performed using a predefined function. Gaussian mixture model, center-based clustering, and similar methods can be used for partition based clustering.

Let us try to understand how incremental clustering would work. In incremental clustering, first and foremost the requirement of the number of clusters prior to clustering does not persist. Here the clustering takes place on the basis of threshold value. Proper selection of threshold value results in good-quality clusters. The outliers are also handled at the same time, depending on the pattern of the data; new cluster formulation also takes place. There is always a condition that the clusters should be well separated from each other. This separation between the clusters is generally on the basis of the distance as the proximity. This distance can be Euclidean distance, Manhattan distance, or any other mechanism to determine decision-centric similarity between two or more data points. One can even use closeness among the set of data or data series for determining the similarity between two data series. With incremental clustering, the clusters are refined incrementally when new data points are added. It is

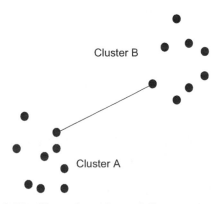

Figure 8.15 Cluster formation and distance measurement.

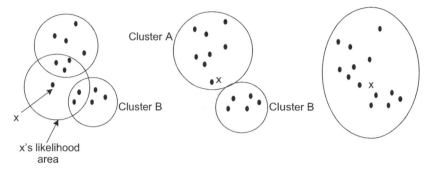

Figure 8.16   Handling outliers: merging the clusters.

made sure that the bifurcation is done properly. There can be overlap among one or more clusters. The handling of different clusters, finding outliers, and merging of the clusters is depicted in Figure 8.16.

Once the cluster formation takes place, the clusters are represented appropriately. This representation is important, as it will be the knowledge base used in the further stages of incremental learning operation. Furthermore, with a new dataset coming into picture, the clusters are updated incrementally and at the same time merging or discarding of clusters that are dormant will also take place. Figure 8.16 depicts the cluster merging. Since the databases are dynamic and new datasets and data points become available, there is need to cluster incrementally. Let us assume that a simple clustering method is used in the dataset and we obtained k clusters:

$$c = \{c_1, c_2, \ldots, c_k\}, \quad 1 \leq i \leq k$$

Let us assume that because the database is incremental, new data points $y_1, y_2, y_3, \ldots, y_m$ are added in the database over the time. Handling of these points to existing clusters is depicted in Figure 8.17.

The first step is the formation of the cluster. Any two points belong to the same cluster if they show sufficient similarity with a data point in the cluster. This is generally determined using threshold and measures for similarity. With multiple datasets being classified in two clusters, there might be a need to merge them. Figure 8.18 depicts this property resulting in a single cluster formation using multiple clusters. This depends on the threshold and distance among set of data points.

8.5.1   Incremental Clustering: Tasks

Incremental clustering can rightly be defined as clustering that updates clusters incrementally. This focuses on set of affected objects. Clustering takes place incrementally in terms of cluster formulation and updating knowledge with accurate decision making considering the impact of new datasets on the formed clusters. Let us now have a look at what tasks are carried out by incremental clustering, which will give us a clear idea of how the clustering takes place incrementally.

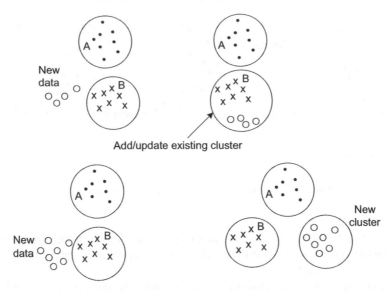

Figure 8.17 Cases of handling new data.

Incremental clustering keeps track of interdependencies, changes in potential membership of clusters. The tasks that are done in incremental clustering are

- Generate new clusters dynamically, independent of cluster number prior to clustering.
- With new unlabeled data, accommodate/classify the data in existing cluster or, if required, form new clusters.

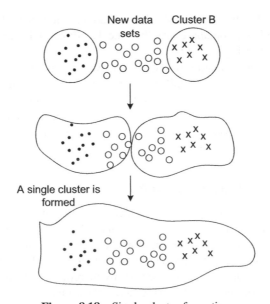

Figure 8.18 Single-cluster formation.

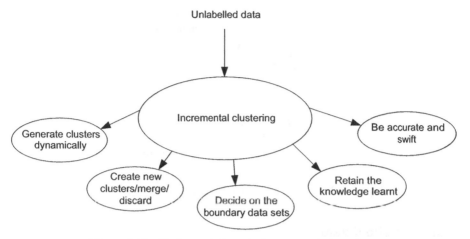

Figure 8.19   Tasks carried out by incremental clustering.

- While formulating clusters, make decisions with regard to merging of clusters or discarding of scenarios that are not required further.
- With the learning taking place incrementally, consider the impact of the data classification in terms of knowledge gained.
- Be accurate and fast in the learning task.

Figure 8.19 summarizes the tasks carried out in incremental clustering.

8.5.2   Incremental Clustering: Methods

Let us now discuss incremental clustering approaches. Incremental clustering can take place in single scan or in some cases takes two scans. Single-scan approaches are faster than two-scan, but the accuracy level is a bit low compared to the two-scan. In clustering as mentioned earlier, the clusters are generated dynamically without the precondition of the cluster number. Since the clustering needs to be accurate, the accuracy of incremental clustering is governed by the threshold value.

Threshold value plays a very critical role during the clustering. There are various approaches to generation of this threshold value. This is dependent on the type of data that needs to be clustered. Incremental clustering methods prefer an approach of tuning of threshold value with different sets of same type of input. With this tuning, the threshold value obtained classifies the data with great accuracy. The other approach is where the threshold value is obtained depending on the frequencies of the attributes. This is where typically the distribution of the data is considered.

To have the dynamic threshold value for clustering, the clustering needs to be intelligent—that is intelligent in terms of the data handled. It should understand the distribution of the data and accordingly be able to adjust/modify its value so as to tune in for better accuracy.

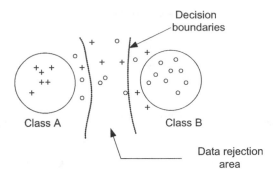

Figure 8.20   Data-rejection area with decision boundaries.

8.5.3   Threshold Value

It becomes a more and more difficult task in the case of new information to identify whether it is a completely new scenario or one similar to what is already learned. These decisions are generally made based on the threshold value. There are a number of strategies while making decisions, and in some cases the threshold is the region that identifies the region of confusion. Another approach is hard threshold. In this case, we try to identify the new instances for learning. While clustering incrementally, one should always be very careful on the threshold value. The threshold value is a key aspect during incremental clustering. All decisions about the clustering are governed by this threshold value. There are many ways for threshold-value calculations. The accurate boundary-region detection can be helpful in this case. This can be done using maximum likelihood. Figure 8.20 below shows the decision boundaries.

The threshold value decision can be made on the following basis:

1. Typically the value can be decided on the distance measure between data series.
2. It can also be the closeness value between two series. The details about this closeness value will be discussed later.

The threshold always is a criterion, which will be used for classification and is generally decided after analyzing the input data pattern.

One of the most important points to address here is—is it possible to have the threshold value updated or changed dynamically with the variation in the input data? This will be the most difficult and critical task to look upon. With the changes in the input data, by changing the threshold, noteworthy results can be obtained.

8.6   SEMISUPERVISED INCREMENTAL LEARNING

Ability to learn from labeled as well as unlabeled data can prove to be very useful for incremental learning. In real-life scenarios, new information is not labeled. To make this unlabeled information play an active role in learning, we need semisupervised learning. Semisupervised incremental learning considers the following aspects:

Identification of relevant unlabeled data that can be mapped with reference to existing training set.

Understanding the relevance of that data and accommodating that data for learning.

With further exploration, keep updating the impact and the relationships dynamically.

In addition to the similarity and closeness-related information used by unsupervised clustering, in many cases a small amount of knowledge is available concerning the group or class of the data. Sometimes it is either pairwise (must-link or cannot-link) constraints between data items or class labels for some items. Instead of simply using this knowledge for the external validation of the results of clustering, one can imagine letting it "guide" on some sort of improvement in clustering policy. Thus the use of controlled supervision with reference to knowledge allowing unlabeled data to take part in reference data is referred as semisupervised clustering. Generally, the available knowledge is incomplete and far from being actually representative knowledge to provide the right classification in real-life scenario. New unlabeled data reveals new information such as

Similarity of new data with existing classes.

Additional properties.

The overall impact of new data on overall cluster formation.

Finding groups of objects such that the objects in a group will be similar (or related) to one another and different from (or unrelated to) the objects in other groups.

In the clustering process, the intracluster distances among objects need to be minimized while intercluster distance need to be maximized. A typical scenario is depicted in Figure 8.21.

Clustering employs different methods and similarity measures. In similarity-adapting methods, an existing clustering algorithm using some similarity measure

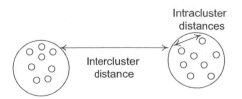

Figure 8.21   Inter- and intracluster distance measurement.

is employed; these measures are based on distance measurement. But the similarity measure is adapted so that the available constraints can be easily satisfied. The similarity measures include Euclidean distance, Mahalanobis distances adjusted by convex optimization, or statistical similarity measures.

In incremental clustering, the clustering algorithm itself is modified or improved so that user-provided constraints or labels can be accommodated and included to bias and to provide the appropriate clustering. This can be done by performing a transitive closure of the constraints and using them to initialize clusters.

Furthermore, semisupervised incremental learning is very important from various real-time learning perspectives. Related points must link while unrelated cannot link. The following aspects are considered in semisupervised learning.

- Large amounts of unlabeled data exists
    More is being produced all the time.
    All new data come in unlabeled form.
- Expensive to generate labels for data
    Usually requires human intervention.
    The producing of labeled data needs experts.
- Use human input to provide labels for some of the data
    Improve existing naive clustering methods.
    Use labeled data to guide clustering of unlabeled data.
    End result is a better clustering of data.
- Potential applications
    Document/word categorization
    Image categorization
    Bioinformatics (gene/protein clustering)

The prior labels are provided by the user in the beginning. This labeling is based on user's knowledge about data points such as which two data points must be linked while which documents should not be linked. This forms the basis for learning. In semisupervised learning, the initial knowledge is further exploited in the light of new data points and new learnings.

With supervised and unsupervised methods discussed, we can get the best of both worlds using semisupervised incremental learning. The data available here are labeled as well as unlabeled. The incremental learning now will happen as a combination of supervised and unsupervised learning.

The tasks looked upon in this case of learning are:

1. From the labeled data available, build knowledge base incrementally.
2. With the unlabeled data, update and restructure the knowledge base incrementally.
3. Make decisions about the new instance on the basis of knowledge base and update.

8.7 INCREMENTAL AND SYSTEMIC LEARNING

We have studied systemic learning in earlier chapters. The point of the argument here is: Is there a relation between incremental learning and systemic learning?. Incremental learning and systemic learning go hand-in-hand. Systemic machine learning needs incremental learning. As the new facets of the system reveal over the time, there is a need to accommodate them in overall knowledge building. As said earlier; incremental learning needs to be active all the time and so does systemic machine learning. So to make decisions, it needs to react as per the current state. It can work as an agent that is up at any instance of time. Patterns or changes in the input depending on the time stamp need to be governed for better outcomes in classification, and accordingly the knowledge needs to be updated. The question here to address is about the rewards. What rewards do we get in incremental learning? The incremental learning needs reward from the system to take actions and to build knowledge. Interestingly, systemic learning considers many perspectives and hence the incremental learning in systemic learning needs to consider different perspectives.

The most important part of systemic learning is systemic knowledge building. As the information comes in bits and pieces, it needs to be combined and without losing the knowledge and perspectives built in the past. Also, there is a need to build a new perspective. The relationship between incremental and systemic learning is depicted in Figure 8.22.

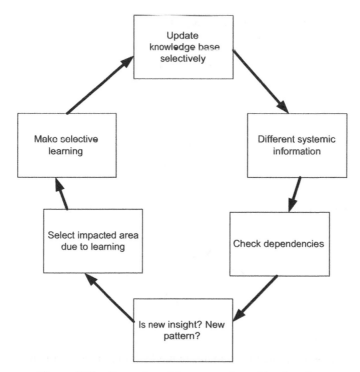

Figure 8.22 Systemic and incremental machine learning.

8.8 INCREMENTAL CLOSENESS VALUE AND LEARNING METHOD

Here we will introduce a new factor for calculation similarity between two series, called the "closeness" factor.

Closeness between the series is calculated with the probabilistic approach. The closeness value is explained as follows.

Consider two data series S_1 and S_2. $S_i(j)$ is point j in series i. $T(j)$ is the total of parameters of series

$$T(j) = \sum_{j=1}^{n} S_1(j) + S_2(j)$$

The probability of outcome S_1 is calculated as

$$P = \frac{\sum\limits_{j=1}^{n} S_1(j)}{\sum\limits_{j=1}^{n} T(j)}$$

The expected value of $S_i(j)$ is calculated

$$S_i(j) = P * T(j)$$

An error $c(j)$ is defined as

$$c(j) = \frac{P \times T(j) - S_i(j)}{\sqrt{T(j) \times P \times (1-P)}}$$

Finally, the closeness "C" between the series is calculated as

$$C^2 = \frac{\sum\limits_{j=1}^{n} c(j)^2 \times w(j)}{\sum\limits_{j=1}^{n} w(j)}$$

where $w(j) = \sqrt{T(j)}$.

Use of the closeness factor for incremental learning gives results that are worth noticing.

8.8.1 Approach 1 for Incremental Learning

Consider a cluster C_1.

Let C_1 have n data series $D_1, D_2, D_3, \ldots, D_n$.

$$D_1 = \{e_{11}, e_{12}, \ldots e_{1m}\}$$
$$D_2 = \{e_{21}, e_{22}, \ldots e_{2m}\}$$
$$D_n = \{e_{n1}, e_{n2}, \ldots e_{nm}\}$$

Each series has m elements. Each cluster has stored sum of all data elements in that cluster with that cluster.

$$\text{sum} = \sum_{i=1}^{m} \sum_{j=1}^{n} e_{ij}$$

This cluster is represented by P values:

$$C_1 = P_1, P_2, P_3, \ldots, P_m$$

Now a new data series or cluster come in to the picture and belong to cluster C_1. The addition of this new cluster will change the pattern. These changes can be represented and made incrementally.

$$N_d = N_{d1}, N_{d2}, N_{d3}, \ldots, N_{dm}$$

$$\text{sum}(N_d) = \sum_{i=1}^{m} N_{di}$$

The new cluster C_{new} is

$$C_{\text{new}} = P_{\text{new1}} + P_{\text{new2}} + P_{\text{new3}} + \cdots + P_{\text{newm}}$$

where

$$P_{\text{NEW}k} = \frac{P_k \left(\sum_{i=1}^{m} \sum_{j=1}^{n} e_{ij} \right) + N_{dk}}{\left(\sum_{i=1}^{m} \sum_{j=1}^{n} e_{ij} \right) + \sum_{l=1}^{m} N_{dl}}$$

$$P_{\text{NEW}k} = \frac{(P_k)(\text{SUM}(C_1)) + N_{dk}}{\text{SUM}(C_1) + \text{SUM}(N_d)}$$

So if the sum is saved with each cluster, then we can modify the clusters incrementally.

8.8.2 Approach 2

(To calculate C, i.e., modified between two clusters when one of the clusters is modified).

This will give us results very close to expected in the case of a stabilized cluster. If the clusters are not stable, it may lead away from correct results. In this case, there is no need to save the sum of all data elements of the clusters.

Consider cluster C_1 to be

$$C_1 = P_1 + P_2 + P_3 + \cdots + P_m$$

Consider new data series ND to be

$$ND = ND_1, ND_2, ND_3, \ldots, ND_n$$

Number of series in a cluster $= n$.

Let the new cluster values after inclusion of this series in a cluster be

$$NP = NP_1, NP_2, NP_3, \ldots, NP_n$$

Then

$$NP_k = \frac{(P_1 m) + \left(D_k / \sum_{i=1}^{n} D_i\right)}{(m+1)}$$

8.8.3 Calculating C Values Incrementally

This method calculates the C value incrementally with reference to distance between two clusters.

We have new data series D that will be included in cluster C_1. Now the C value of this series with respect to each other series will change. Here a proposal is given to track this change incrementally.

There is one proposal that involves slightly less calculations and can give results very close to expected ones.

C_2 is another cluster. The C value of cluster C_2 with respect to C_1 is given below.

$$C_{NEW}^2 = \frac{\sum_{j=1}^{n} \left((C_2(j) - C_1(j))^2 / (\sqrt{C_1(j)} + C_2(j)) \right)}{\sum_{j=1}^{n} \sqrt{C_1(j) + C_2(j)}}$$

Here the new C value is not mentioned in terms of old C. But $C_2(j)$, which is a new jth P value for cluster C_2, can be represented in terms of old P value.

Take series

$$S_1 = e_{11}, e_{12}, e_{13}, \ldots, e_{1n}$$
$$S_2 = e_{21}, e_{22}, e_{23}, \ldots, e_{2n}$$
$$S_3 = e_{31}, e_{32}, e_{33}, \ldots, e_{3n}$$

So, the pattern generated is

Pattern $= (e_{11} + e_{12} + e_{13}), (e_{12} + e_{22} + e_{32}), \ldots, (e_{1n} + e_{2n} + e_{3n})$

P of the pattern

$$= (e_{11} + e_{12} + e_{13})/\sum \text{All}, (e_{12} + e_{22} + e_{32})/\sum \text{All}, \ldots (e_{1n} + e_{2n} + e_{3n})/\sum \text{All}.$$

Now we will calculate $c_i(p_i)$. To calculate $c_i(S_1, \text{pattern})$ we will calculate $p(S_1, \text{pattern})$

$$p(S_1, \text{pattern}) = \frac{\sum\limits_{i=1}^{n} S_{1i}}{\sum \text{All}}$$

$$c(j) = \frac{p \times T(j) - S_i(j)}{\sqrt{T(j) \times p \times (1-p)}}$$

After calculating in the same way for S_2 and S_3
We will get
$c_i(S_1, \text{pattern}) = c_{11}, c_{12}, \ldots, c_{1n}$
$c_i(S_2, \text{pattern}) = c_{21}, c_{22}, \ldots, c_{2n}$
$c_i(S_3, \text{pattern}) = c_{31}, c_{32}, \ldots, c_{3n}$.
The overall C series for the cluster can be represented as weighted average of these three elements.

$$C^2 = \frac{\sum\limits_{j=1}^{3} c(j)^2 \times w(j)}{\sum\limits_{j=1}^{3} w(j)}$$

Here we can take the weight equal to the sum of series where the weight function is $w(j) = \sum S_j$.
Now suppose we have C for three series, and now series S_4 is added to the same cluster. This will modify P_i values, and this is given by the equation discussed above.
For the jth element, the number of series is i.
These equations are for jth element in the C series. In the same fashion, the remaining elements can be calculated.

$$C^2_{\text{New}j} = \frac{\sum_{i=1}^{n+1} c(i)_j^2 \, W(i)_j}{\sum_{i=1}^{n+1} W(i)_j}$$

$W(i)_j$ is e_{ij}, that is, the jth element in the ith series is expressed as

$$\frac{\left(\sum_{i=1}^{n} c(i)^2 \times W(i)_j\right) \Big/ \sum_{i=1}^{n+1} W(i)_j + (c(n+1)^2 \times W(n+1)_j)}{\sum_{i=1}^{n+1} W(i)_j}$$

$$= \left(\left(\sum_{i=1}^{n} c(i)^2 \times W(i)\right) \Big/ \sum_{i=1}^{n} W(i)\right) \times \left(\sum_{i=1}^{n} W(i) \Big/ \sum_{i=1}^{n+1} W(i)\right) + \left(c(n+1)^2 \times W(n+1)\right) \Big/ \sum_{i=1}^{n+1} W(i)$$

$$= (C_{\text{OLD}})^2 \times \left(\sum_{i=1}^{n} W(i) \Big/ \sum_{i=1}^{n+1} W(i)\right) + \left(c(n+1)^2 \times W(n+1)\right) \Big/ \sum_{i=1}^{n+1} W(i)$$

$$= \frac{(C_{\text{OLD}})^2 \times (\text{Weight}_{\text{OLD}})}{(\text{Weight}_{\text{NEW}})} + \frac{\left(c(n+1)^2 \times W(n+1)\right)}{(\text{Weight}_{\text{NEW}})}$$

Here Weight$_{\text{(old)}}$ for the jth element is the sum of all jth elements for n series.

$$\text{Weight}_{\text{old}} = e_{1j} + e_{2j} + \cdots + e_{nj}$$

Weight$_{\text{(NEW)}}$ for the jth element is the sum of all jth elements in $n+1$ series:

$$\text{Weight}_{\text{new}} = e_{1j} + e_{2j} + \cdots + e_{(n+1)j}$$

$c(n+1)$ is c value for the jth element in $n+1$ series.
$W(n+1)$ is the jth element in $n+1$ series.
In the same fashion, the rest of the elements in the series can be calculated.

Thus we can get C_{new} in terms of C_{old}. We know old weight and new weight. We can calculate c of new series with respect to the pattern. As we store the sum of all the series in a cluster, old weight and new weight are known to us. $W(n+1)$ is the sum of the new series and is calculated very easily and so is $c(n+1)$, that is, c between pattern and new series.

$$\text{Overall } C(\text{NEW}) = \frac{\sum_{i=1}^{n+1} \sqrt{P_i(\text{NEW}) \times C_i(\text{NEW})^2}}{\sum_{i=1}^{n+1} \sqrt{P_i(\text{NEW})}}$$

8.9   LEARNING AND DECISION-MAKING MODEL

Figure 8.23 gives the architecture of the new forecasting module. Its applications vary from health-care decision making to hospitality industry and revenue management. This forecasting tool will lie beneath the decision system.

Complete decision-system architecture for decision making based on incremental learning is depicted in Figure 8.24. The decision manager is responsible for decision making and works on historical data and behavior mapping. Qualitative inputs and incremental quantitative inputs facilitate decision making.

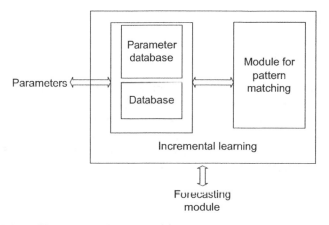

Figure 8.23   Incremental learning and forecasting.

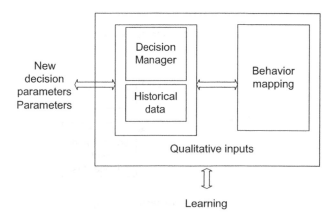

Figure 8.24   Incremental learning and decision making.

8.10   INCREMENTAL CLASSIFICATION TECHNIQUES

Learning results from classification. This could be classification of document types, text, objects, or problem at hand. Hence incremental classification is an important part of incremental learning. The incremental classification allows accommodating new data points for classification without learning from scratch. Let us take an example of document classification. Let us assume that a classifier is trained to classify the sports news and political news. Sports news are moved to the last page of the newspaper while political news are put on the first and second page of the newspaper based on priority. Let us assume that a new type of news comes into picture say scientific news and scientific sport news. In light of these new types we have four types of news:

Sports news

Political news

Science news

Sport-science news

With two additional types of news, one option is learning from scratch and building a classification system that can classify into four classes. The other option is allowing incremental learning, where learning with reference to political news remains untouched. Even the old class of sport news is updated incrementally and two new classes—science and sport science news—are introduced. Figure 8.25 shows the classification—with and without incremental learning.

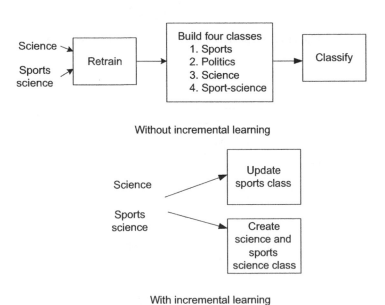

Without incremental learning

With incremental learning

Figure 8.25   Example: with and without incremental learning.

8.11   CASE STUDY: INCREMENTAL DOCUMENT CLASSIFICATION

Various documents need to be classified for any automated document-management system. This may include forms, document images, pictures, and unstructured documents. With information explosion, more and more documents become available and become part of repositories. These documents may belong to already trained classes or may be a completely new type of document. Slowly there are a large number of classes, huge training sets, and relationships of different types among different documents. This imposes a strong need of classification of documents incrementally not only to save the time required but also to retain knowledge built in the past. Another advantage of incremental classifier is that it can start classifying the document without training. This is depicted in Figure 8.26.

Let us take an example of mortgage document classification. This typical application may include different loan forms, notes, riders and so on. Let us assume that it includes forms 1003 and 1004 and an adjustable rate note. These forms are classified based on the term frequency. Now a new document deed of trust (DOT) is introduced. Since there is no area of impact due to DOT, the feature vectors can just be added and incremental learning can result in knowledge retention as well, since there is no need the system to learn from scratch.

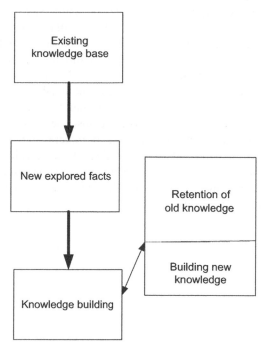

Figure 8.26   Incremental document classification and knowledge management.

Initial Trained Set→ Classification include classes ARN, FRN, and 1003 and 1004
New documents and classes—DOT
Incremental accommodating new classes—weight calculation—new cluster is formed
Formation—knowledge building and document management

The closeness between new_document and ith class can be represented as

$$\text{Closeness\_}i = \text{closeness}(\text{new\_document}, \text{class\_}i)$$

The learning takes place on the training set with all classes where closeness is below threshold and the new document.

8.12 SUMMARY

Incremental learning is the next big step in terms of decision making. Having a strong decision-making capability and taking into account the results from the industry perspective, it is worth noting. With the learning active all the time, incremental learning methods have great potential for forecasting. The decisions inferred are on analysis of the data pattern and with accurate classification results; the proposed changes in existing system by the learning method will have a great impact on the productivity. Incremental learning can be viewed as one of the most important aspects of systemic learning. The knowledge built in every phase and from different perspectives allows building systemic view and facilitating incremental learning.

Semisupervised learning allows us to learn from the unlabeled data along with labeled data. This can allow choosing the relevant unlabeled data for learning. Incremental semisupervised learning allows us to build learning parameters while retaining the useful knowledge built in past. The absolute incremental learning can lead to many different issues. The preferred way of incremental learning is selective incremental learning. Here the selective updating of old learning parameters and building new feature vectors take place. This allows knowledge refinement and knowledge building. Incremental clustering allows building clusters based on similarity and distribution of new revealed facts.

CHAPTER 9

Knowledge Augmentation: A Machine Learning Perspective

9.1 INTRODUCTION

The purpose of any type of learning is building knowledge and managing the knowledge to exhibit superior decision making. This involves use of right knowledge at the right place and at an appropriate time. New scenarios and new information keep coming in all real-life problems. Hence the knowledge built needs to be augmented and used effectively for future knowledge building. The objective of machine learning is to make machines learn to augment knowledge and further build an ability to improve knowledge and use the knowledge effectively for decision making. This is a continuous process, and learning is an important aspect of a knowledge life cycle. This empowerment allows the system not only to perform in a similar scenario more effectively next time but also to perform in a smart way in case of new scenarios.

Knowledge acquisition is the process of identifying new relevant information, absorbing the information, and storing it. This information is stored in the memory and can be later retrieved. The process of sorting, mining, storing, and retrieving relevant information depends heavily on the information storage, organization, and representation. Knowledge acquisition can be improved through better learning and mapping. It can be improved by considering the context and relationships of information. This includes understanding the purpose and function of the desired information along with its relationship. Knowledge acquisition becomes more efficient when the learner focuses on the meaning of the new material and also the holistic relationship of this information. For successful knowledge acquisition, systemic dependencies need to be considered while learning. The success of comprehension and the ability to understand, manage, and improve learning can help in improving the knowledge acquisition. Knowledge acquisition includes the building, elicitation, collection, analysis, modeling, and validation of knowledge for knowledge engineering and knowledge-management projects.

Reinforcement and Systemic Machine Learning for Decision Making, First Edition. Parag Kulkarni.
© 2012 by the Institute of Electrical and Electronics Engineers, Inc.
Published 2012 by John Wiley & Sons, Inc.

One of the most important aspects of knowledge building and reuse remains the context. Context and knowledge acquisition go hand-in-hand. Knowledge capture involves various parameters such as relevance and time dimensions and has very broad scope. It also needs to provide means for acquiring knowledge from different sources in the system. Knowledge acquisition takes place in every learning phase. It can take place based on learning from images, objects, data, or patterns. Many machine learning approaches offer human-guided knowledge acquisition. The first important aspect of knowledge acquisition is to focus on relevant features, and hence there is a need to identify the relevant features. The second part is about understanding these features and knowing the behavior of these features, that is, understanding rules and relationships of features with reference to categories of interest. Another most important aspect is building knowledge in a dynamic scenario with the ability to learn incrementally. The collaborative assessment of knowledge with the reference application and knowledge building in multiagent systems is also discussed in this chapter. The generalization of the knowledge coming from multiple sources, along with using this knowledge for decision making, needs multiperspective learning and decision making. Collaboration and competition among agents can be used for knowledge building and augmentation.

Knowledge augmentation is about building and providing value-added information on top of the existing knowledge with use of metadata and in light of revelation of new facts and availability of new information. It involves knowledge mining, knowledge integration, incremental knowledge building, and new knowledge representation. Figure 9.1 shows the basic view of knowledge augmentation. The knowledge cycle includes capture, preservation, augmentation and dissemination, and use of knowledge. The knowledge cycle is depicted in Figure 9.2.

Explicit feedback mechanisms, adaptive learning, and systemic perspectives should be in place to allow these targeted uses of the knowledge to inform and guide the acquisition of the knowledge. Further multiperspective learning is required for selecting appropriate knowledge-acquisition methods. In this process, the most important part is using the knowledge that is built already more effectively and

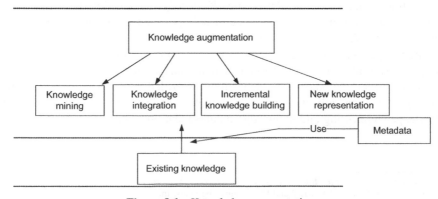

Figure 9.1 Knowledge augmentation.

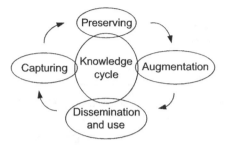

Figure 9.2   Knowledge cycle.

correcting the hypothesis if required. There are two aspects of this knowledge base: (1) Some part of the knowledge can be used as it is. It is rather the knowledge where changing scenarios rarely impact on it. (2) In the next level, the knowledge needs to be refined in light of the new facts. The multidisciplinary concept integration can be used for knowledge building.

This chapter describes the entire knowledge life cycle and the process of learning to the phase of accretion with reference to different machine learning approaches. Furthermore, the knowledge needs to be represented in such a way that the learning system can reuse it for future learning and decision making. We will discuss these knowledge-building aspects with reference to machine learning methods.

9.2   BRIEF HISTORY AND RELATED WORK

Around the 6th century BC, various scientists and researchers from different countries including Greece, India, and Russia learned the need for the development of reasoning-based methods and techniques that will make knowledge acquirement easy. This knowledge acquirement needs to be based on empirical information of that era. According to those researchers to acquire knowledge, it always needs to be based on previous historical information, which in turn helps learning for estimation and analysis. The paradigm of purely historical knowledge-based learning has its own limitations. The first few academicians to challenge rational reasoning were Plato, Socrates, and Democritus from the 4th century BC. Aristotle (384–322 BC) is a well-known founder of academic formal logic. Epistemology is the term related to the system of techniques and tools that facilitate generation of scientific knowledge and scientific relationships-based learning. Starting from inductive methods, rational reasoning, and mathematical logic as topics useful for acquisition of knowledge, today's era needs systems to have the capability of building and augmentation of knowledge. The system should behave smartly even in the case of similar, not so similar, complex, and not so complex scenarios. To build the set of required systems, let us first understand what knowledge is all about: It is not just the storage of a large amount of related information useful for estimation, forecasting, analysis. It is rather a useful contextual interpretation of this information that can help to achieve the objective. Today this storage and retrieval of information is done with the help of

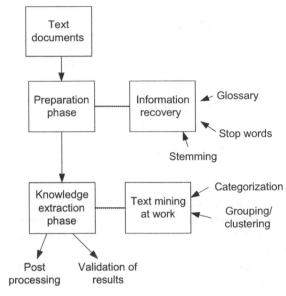

Figure 9.3   Knowledge discovery process from text.

latest artificial intelligence, business intelligence, and data-mining-based systems/ application software. Knowledge can further be viewed as information known to a person to complete the potential problem of completeness. Figure 9.3 depicts a typical knowledge discovery process from text.

J. H. Johnson, P. D. Picton, and N. J. Hallam, in their research paper in 1994 titled "Safety-critical neural computing: Explanation and verification in knowledge augmented neural networks," addressed the problem that neural networks, especially conventional ones, cannot contain *a priori* knowledge, and it is not possible to explain their output [1]. "Black-box classifiers" theory is applicable to architecture of neural classifiers. By themselves, neural networks cannot communicate with human decision makers in human terms. Hence, knowledge augmentation plays an important role in neural networks. In the case of applications such as software engineering where every project is different and varies in complexity, simple pattern-based historical-information-based learning shows many limitations. The knowledge building based on exploration and experience-based correction proves to be useful in the case of these complex problems.

In the research article by Park, Yu, and Wang [2], they have proposed a method to enhance the performance of knowledge-based decision-support system. The qualitative reasoning (QR) approach can be used to augment knowledge in the dynamic scenario when the knowledge is incomplete and the scenarios are changing continuously. The knowledge base needs to be updated with every new finding. The problem of completeness of the knowledge base may result in the case of static knowledge base. The dynamic knowledge base with appropriate knowledge augmentation strategy as well as method can help to overcome this problem.

In the year 2002, the research paper [3] proposed a new idea: Structured documents have different objects, and the contents are mapped to these objects. The effective retrieval of structured documents requires a content-based retrieval of objects that takes into account their logical structure. This paper proposes a formal model that representing structured documents where content can be viewed as knowledge contained in objects and the logical structure can be captured by a process of knowledge augmentation. Structurally connected objects can help to augment the knowledge.

In the year 2004, Richard Dazeley and Byeong-Ho Kang introduced an augmented hybrid system referred to as MCDR [4]. It uses Multiple Classification Ripple Down Rules (MCDR), a simple and effective knowledge-acquisition technique, combined with a neural network. The authors have performed research activities to identify keywords or groups of words which help in the incremental knowledge acquisition, in their paper titled "An Augmentation Hybrid System for Document Classification and Rating." The keywords are not enough to get the context, and they suffer from limitations in the real-life scenario where knowledge is incomplete in the absence of context.

The following details about knowledge were mentioned in Ref. 5. To gain access to this valuable information, discussion search is concerned with retrieving relevant annotations and comments with respect to a given query, making it an important means to satisfy users' information needs. Discussion search methods can make use of a variety of context information given by the structure of discussion threads. Evaluation shows the suitability of knowledge augmentation for the task at hand. This paper discusses the evaluation of knowledge augmentation in discussion search.

Bodenreider and Zang [6] mentioned that the objective of their study is to evaluate the contribution to semantic integration of the semantic relations extracted from concept names, representing augmented knowledge. It investigated multiple augmentation methods including reification, nominal modification, and prepositional attachment.

In a recent study in 2009, Chen, Jhing-Fa Wang, and Jia-Ching Wang from the Department of Electrical Engineering, National Cheng Kung University, Tainan, Taiwan mentioned that a video knowledge browsing system, which can establish the framework of a video based on its summarized contents can expand them by using online correlated media [7]. Thus, users can not only browse key points of a video efficiently but also focus on what they are interested in. In order to construct the fundamental system, they make use of their previous proposed approaches of transforming a video into a graph. After the relational graph is built up, the social network analysis is then performed to explore online relevant resources. They also apply the Markov clustering algorithm to enhance the results of the network analysis.

When we want a machine to exhibit an intelligent behavior, it should be able to augment the knowledge. Here systemic knowledge augmentation is to understand the features, build knowledge, and represent it from systemic perspective. This is like understanding the relationships among different pieces of knowledge and building a systemic knowledge perspective that can be used in different scenarios. More than one perspective need to be combined. A typical multiagent architecture for knowledge building is depicted in Figure 9.4.

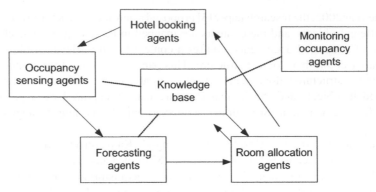

Figure 9.4   Multiagent architecture for knowledge building.

The learning system development with multiagent architecture is depicted in Figure 9.5. In the first phase, systemic goal derives the overall knowledge augmentation process. The performance measures need to be finalized for exploration. Adaptive learning allows selection of the learning policy and learner based on context and decision scenario with reference to exploration and measurement of impact of different actions and outcomes. Agents learn about these outcomes through exploration, while the impact is observed using different parameters across the system. The multiple agents learn for the same decision scenario to build the context.

After learning about the researchers' activities in varied domains for augmentation of knowledge, let's learn various information-gathering details, including knowledge acquisition and elicitation followed by entire life cycle of knowledge related to various case studies. Knowledge and relevance are both related in a way that relevance

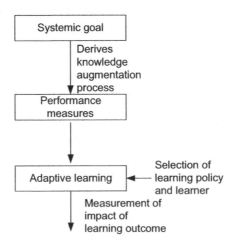

Figure 9.5   Learning system development with multiagent architecture.

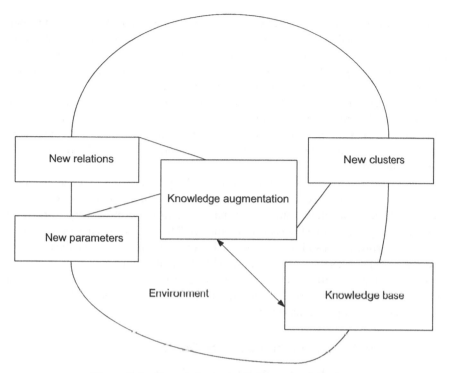

Figure 9.6   Knowledge and relevance augmentation.

helps us to build knowledge. Knowledge and relevance augmentation in light of new explored facts is illustrated in Figure 9.6. Knowledge is augmented with new parameters, and relations are augmented with reference to the existing knowledge base.

9.3   KNOWLEDGE AUGMENTATION AND KNOWLEDGE ELICITATION

Knowledge can be generated based on gathered information and a chronological/ pragmatic and practical database. Knowledge elicitation is a major stage to gather information. It is the act of bringing or drawing out something embryonic. It is defined as the method to arrive at a truth by logic. Knowledge-elicitation methods are strategy based, goal based, or process based. This information can come in different forms and from multiple sources. The details of all individual methods are given below.

9.3.1   Knowledge Elicitation by Strategy Used

This method of information acquisition is based on the various problem-related techniques, and they are discussed below.

In the case of software development projects, the client provides a problem statement first. Based on this given problem statement, it is essential for the analyst to suggest feasible solutions in the form of various requirements (technical and practical]. To analyze these hidden requirements from a client, it is essential to understand the given problem statement unambiguously.

Experts need to solve a given problem either using their experience or using various applicable methods including discussion, classification, or analysis.

Problem discussion involves the discussion of what was the cause of the current situation, what has happened, and what are the feasible and quick effective solutions to resolve the given difficulty.

Problem analysis involves decomposition of the problem into realistic classes to decide the strategy to apply. To analyze the given problem, the analyst may make use of various IT tools available online and make sure to reuse previously worked out solutions (if any) to resolve the issue immediately. The analyst gets a good amount of help to learn from various IT systems/tools. For example: root cause analysis (RCA), mind tools (MT), cause analysis tools (CAT), and many more.

The classification process contains certain criteria to divide collected data and store it in different databases. The learning policy defines the context for classification, decision making, and learning.

9.3.2   Knowledge Elicitation Based on Goals

This method describes how important it is to understand the learning and decision goals and objectives before prescribing any solution or analyzing a given problem. In this case, the knowledge acquisition is goal based. The learning system gathers information by focusing on the current goal and keeps track of business logic and market conditions while processing further knowledge acquisition.

Learning systems that use multiple information sources and cooperative information gathering refer (if need be) to the objectives, set short-term/long-term goals of the system in a specific decision scenario, divide the set of processes (to acquire knowledge) into various categories, and act based on assigned priorities. Goals are classified into various categories, so as to focus on a small portion at a time, for successful acquisition of knowledge. If required further, decision-tree-based systems/laddering concepts are fruitful in knowledge elicitation by goal-based method.

9.3.3   Knowledge Elicitation Based on Process

This technique describes the procedure to elicit information.

1. *Through Responses Based on Queries*: This phase involves one-to-one questionnaires to solve a particular problem. Interactions can be direct as well as indirect.
2. *Collecting Artifacts*: This phase collects all information from all the available information sources related to the problem. Artifacts are generally repositories, documents, and similar information sources and urls.

3. *Protocol Analysis*: This phase is based on a rule or procedure to analyze the problem. It involves detailed analysis of the problem and finding a solution to it part by part.

4. *Relationships*: Establishing relations between the problem and the solution obtained is considered here. These relationships can be derived in a statistical way. The relationships can be based on a particular criteria with reference to context and decision scenario.

5. *Observation*: Observation is based on the results obtained from previous experience. This helps to avoid mistakes or errors that occurred earlier. There is typically a knowledge base that can be used for the exploration and deciding the action in new and similar scenarios.

9.4 LIFE CYCLE OF KNOWLEDGE

Data are used to get meaningful information, and further knowledge is built. Data go through many stages in knowledge-building process. Knowledge building is closely associated with learning. Knowledge building results through learning and contextual associations. This happens as learning provides the most required contextual details with reference to the decision scenario. The context allows augmenting knowledge with reference to the new revealed facts. Figure 9.7 depicts knowledge life cycle.

To acquire knowledge, its life cycle necessitates going through various stages using various methods, techniques, and tools. These stages include understanding the context of industry first, followed by gathering raw data and generating/acquiring knowledge along with the most important part, which is sharing this knowledge,

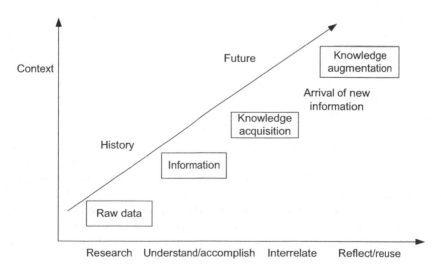

Figure 9.7   Knowledge life cycle: from raw data to augmentation.

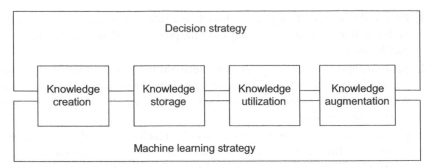

Figure 9.8   Knowledge augmentation based on business and machine learning strategies.

reusing knowledge, and augmenting knowledge. To achieve all these stages of knowledge life cycle it is necessary to make use of various machine learning/technical methods and tools/techniques.

One of the important "advanced machine learning" paradigms is referred to as incremental cooperative machine learning. It includes "incremental clustering." "Incremental clustering" helps to update clusters of knowledge with latest information, which will be extremely essential for managers and decision makers to form effective strategies or estimations. Knowledge augmentation in machine learning results in strategies such as cooperative learning, adaptive learning, and, most importantly, incremental machine learning. Knowledge augmentation keeps the useful built knowledge intact and builds further knowledge on top of it.

As shown in Figure 9.8, the life cycle of knowledge is divided into two major components, which are "decision strategy" and "machine learning strategy." The decision component consists of decision space, decision components, standards and guidelines, policies, and sociocultural environment. The "machine learning strategy" consists of learning policies, intelligent components, tools, and techniques to implement effective augmentation of knowledge which provides values to its society, business, employees, customers, and partners. The tools can be further classified into knowledge creation, knowledge store, knowledge utilization, and augmentation.

There are different categories of knowledge: domain specific, environment specific, contextual, and decision specific. Domain-specific knowledge describes knowledge related to a particular field or domain. For example, in a software domain, knowledge is generated with the help of various applications, databases, patterns, components, and so on. The environment-specific knowledge is more systemic in nature and is about the environment or decision space. The contextual knowledge is more about decision context and is built in decision space with reference to decision scenarios. The decision-specific knowledge depends on decision outcomes and impact. For example, let us assume that there are two types of users' categories: novice and expert. So, a novice user has only introductory knowledge while an expert user has detailed and advanced knowledge on a particular subject.

9.4.1   Knowledge Levels

The knowledge is augmented, and this is a continuous and incremental process. The knowledge is built on basic knowledge where the basic information about decision scenario is built. This is simple pattern-based knowledge building and inference where complex dependencies are not involved. The next-level knowledge is built on that with reference to relationships, mapping, and dependencies. The advanced level or decision-level systemic knowledge includes the knowledge mentioned above along with the impact analysis of different actions in decision space with reference to the systemic context. Let us discuss now in what ways the knowledge is generated along with different levels of knowledge building.

9.4.2   Direct Knowledge

The collected information with direct interaction is verified against hypothesis to build direct knowledge. The immediate, actual information gathered by use of various sensors or by various agents/intelligent agents including seeing, hearing, feeling, smelling, and tasting will be helpful in gathering direct knowledge. The majority of sensors give the same output except for a few cases where, for example, a person with color blindness will observe different colors. The data related to experience that can be experienced by an observer or the system comes under direct knowledge. Direct knowledge builds a platform for building advanced knowledge.

9.4.3   Indirect Knowledge

The learning is based on inference, and deriving knowledge from direct knowledge can be referred as indirect knowledge. Ambiguity can be one of the major problems with indirect knowledge, because this type of knowledge acquisition is based on how the knowledge is comprehended and what is the set of processes used to produce this knowledge including thinking, understanding, and so on. Hence it is said that indirect knowledge is knowledge acquired by processing information. The indirect knowledge is sensitive to learning algorithms.

It also includes intuitive knowledge. In typical human context, it is knowledge stored in subsystem memory that is not accessible to the higher brain functions. In the case of machine learning, it is more an inferred fact and is based on continuous inference and context-based findings that are not directly visible.

9.4.4   Procedural Knowledge

Procedural knowledge is like walking through any algorithm, where how-to instructions are given that tell step-by-step how to do something. Procedural knowledge is a series of instructions to perform a particular task or activity. Procedures are a sequence of percepts or sequence knowledge events that can help in building knowledge spaced out in time. This knowledge results through a series of explorations and combination of individual sequential actions. Procedural knowledge typically helps in understanding sequence of actions leading to desired outcome. This knowledge has various aspects such as measuring intermediate results, understanding the sequence, and importance of performing actions in sequence.

9.4.5 Questions

Questions or queries will help generate knowledge. This type of knowledge can be built based on response of environment to different actions. The questions are a sort of knowledge gaps. Questions in learning are the actions for which response is not known. Questions are part of the knowledge-acquisition process and it is the initiation of the process to find out unknowns. For example, while using search engines online, we may ask a question or enter a keyword for which we seek complete information. Until today, only text-based search engines were available online—furthermore, there are image-based searches, relationship-based searches, advanced searches, filtering searches, and Boolean searches—these are the questions to bridge knowledge gaps. Questions can help in procedural knowledge building and inferential knowledge building.

Asking question to search engines such as Google and Clusty shows completely different results because of various aspects. The background of data mining, along with the concepts of clustering and machine learning, plays a very vital role.

9.4.6 Decisions

Decisions are the directives for the action. The decision question is typically to select among the choices and procedural knowledge that can lead to decision. The conditional selection among the alternatives can cause the procedure to change with respect to the answer to the question. Knowledge of decisions can be a historical record of why certain procedures were executed in a certain way. The decisions are used for learning along with the impact of decisions.

9.4.7 Knowledge Life Cycle

The knowledge life cycle is closely associated with learning, and it is about building knowledge, validating it. Furthermore, there is a need to rebuild the knowledge in new scenarios and in light with new facts. "Knowledge-augmentation life cycle" is a never-ending path that increases with every added information/contents. Knowledge-augmentation life cycle goes through the following phases:

1. *Identification of Need/Understand the Context*: To acquire knowledge, the complete industry details, domain details, decision, and learning strategies must be known related to all contexts. To make use of data-generation tools and techniques, it is essential to gather the empirical raw form of data first. Then based on the requirement related to the type of learning, use of various clustering techniques will be useful. The domain and the related information helps to build the context.

2. *Gathering Information/Acquire Knowledge*: This phase involves collecting of data from different information sources including various knowledge sources such as experts experience, research papers, books, websites, knowledge repositories, and machine learning algorithms including AI, BI, and so on. These information sources are selected based on context.

3. *Analysis of Information*: By applying the various suitable basic analysis methods (these may include clustering, classification, statistical ranking), raw data need to be stored in various relevant groups and mapped to the priorities and decision scenarios. These relevant groups or clusters will then be used to generate behavioral patterns or can be used to infer the behavior. The pictorial information in the form of patterns will be handier for the analyst to make an effective decision, as compared to clustered data in the form of figures or text. To perform proper analysis, it may be essential to combine or divide the formed clusters.

4. *Learning*: The knowledge with the system is in the form of knowledge base. In whole-system learning, the learning system should make the best use of all available data sources and algorithms. Along with the exploitation of the knowledge the system possesses, exploration of the new scenarios and actions makes it possible to learn from experience. The concept of learning is using empirical data-based knowledge, while the system keeps exploring in the case of new scenarios. Let us take an example of a software development company. It has developed a software system for a particular bank successfully. Because of this success, another finance company approaches the same software development organization for developing a system for them. Now if the knowledge built and bank software details are preserved neatly, using data-mining techniques, the entire knowledge can be retrieved as and when required and reused for the new software development as the domains are similar. When the domains remain similar, the protocols, the products, and customers categories are mostly repeating. Hence reuse of knowledge or learning based on acquired knowledge can be easily achieved using advanced machine learning techniques. But in the case of other applications, it would be a combination of exploitation and exploration. In the real-life scenarios, two problems are not same but can be similar or appear to be similar. Learning based on these similarities, differences through exploration can help in augmenting knowledge.Once this software company achieves a name in the finance sector and many other clients from the finance domain approach them for their application system development, then incremental learning needs to be implemented.

    This phase involves storage as well as retrieval of information. The storage involves extensive use of latest database, network technologies, and superior servers with latest pattern generation, pattern-matching algorithms for effective utilization of augmented knowledge. The retrieval process involves the extraction of stored information, as and when required.

5. *Enhancement*: This phase involves polishing and expanding the knowledge. This phase goes on in increasing the knowledge database. This phase is also called KNOWLEDGE AUGMENTATION.

As mentioned above in point no. 4 related to learning and explained in the previous section, the raw data or available information will be stored in the form of basic clusters initially. Technically, on the arrival of new information, it is essential to update these clusters and mapping and relationships with new information. Updates related to these relationships are mandatory to give effectual estimation. Now the question arises: which cluster to update when and also when to formulate new relationships. When the basic clusters are formed, a few more cluster details including representative time/data series, center of cluster, distance between clusters, and threshold range of clusters are also stored. The new information features will be compared with the stored/available cluster details, and the decision is made to update a particular cluster with the new information. If the feature set of already formed cluster does not match with the new information, then the new cluster needs to be generated and these entire technicalities will be neatly taken care of by the Closeness Factor-Based Algorithm (CFBA), COBWEB, or others incremental clustering algorithms. The speed, complexity, scalability, memory utilization, and other technical details vary from method to method.

9.5 INCREMENTAL KNOWLEDGE REPRESENTATION

As discussed in previous sections, the new knowledge never becomes available at once and becomes available in bits and pieces. This new knowledge for effective knowledge building needs incremental knowledge building and representation. The effective knowledge augmentation results through cooperative learning. The incremental learning model allows deriving the relationship of new facts with reference to the knowledge base with the system. The new information is acquired, and with reference to adaptive features and the decision scenario the overall relationship is reorientated in the form of reasoning. Figure 9.9 depicts the information flow and knowledge representation.

Observation of the behavior of the user and environment can help in building training examples. The algorithms perform the acquisition of knowledge on these training examples. The learning about the environment continuously helps in building incremental knowledge representation. The background knowledge is typically reflected in training examples. The experiences need to be represented in the form of knowledge to make future learning more efficient. The incremental knowledge building and representation is about the new relationships, new clusters, and new information built without impacting on relevant, useful knowledge built in the past.

Let us assume that we have three clusters (A, B, C) of 10, 11, and 15 data points each. Each of these clusters represents typical parameters associated with a particular disease. Assume we come across three more data points that represent similar behavior as cluster A. But as we explored with action in decision space, the outcome was different. This new knowledge is built, and it may impact a few data points in cluster A while the knowledge in clusters B and C is intact.

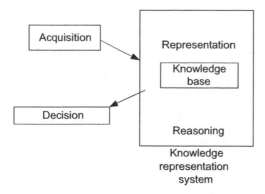

Figure 9.9   Information flow and knowledge representation.

The exploration-based knowledge-augmentation architecture is depicted in Figure 9.10. Here machine learning algorithms model the information, build groups, and determine relevance. This interacts with the knowledge representation. The observations and decisions allow the exploration and learning. This learning is based on assumptions. The interaction with the environment allows the refining of these assumptions. The observations and the impact of decisions is input for knowledge augmentation. New knowledge is represented with reference to the existing groups. The new groups and decision mapping is formed, and even the assumptions are amended in case it is required.

Figure 9.11 depicts incremental knowledge representation. Here knowledge representation and inference module helps to represent the new knowledge scenarios. Data-acquisition system collects data. With inference and perception, the knowledge is represented so that it can be used incrementally for learning.

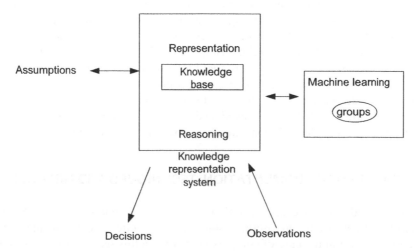

Figure 9.10   Exploration-based knowledge augmentation.

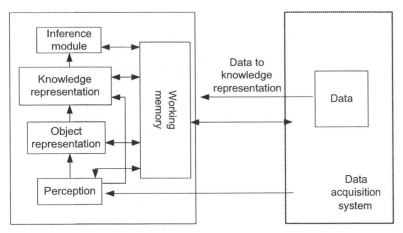

Figure 9.11   Knowledge representation: incremental approach.

9.6   CASE-BASED LEARNING AND LEARNING WITH REFERENCE TO KNOWLEDGE LOSS

In the case of incremental learning and data transformation, there is always a possibility of knowledge loss. This knowledge loss may result due to incorrect context or absence of a complete view. Here the knowledge is built with the assumption that similar problems have similar solutions. Even many learning hypothesis are based on this assumption. When the system comes across a new problem, it tries to apply the knowledge it has. The completely new problem creates an opportunity for learning and a new case for learning. Figure 9.12 depicts problem-based learning, knowledge capture, and knowledge reuse. The knowledge selection and knowledge application based on cases or problems is used to build a knowledge base in case-based learning. Case-based learning is knowledge intensive and tries to build knowledge based on case. It tries to extract parameters from experience and infer based on relationships. Even in a sequence of experiences, knowledge is reused and built to be reused for the future. It tries to build mapping between heuristic rules and solved cases, experiences, and so on.

Figure 9.13 depicts retrieval, reuse, and revisal of knowledge. With reference to a new problem, the knowledge is retrieved; and in the case of applicability, the existing knowledge is reused. The solutions are also retrieved and reviewed.

9.7   KNOWLEDGE AUGMENTATION: TECHNIQUES AND METHODS

Knowledge and relevance augmentation is part of continuous learning. Simple content-based approaches are useful in some cases, but inference-based approaches prove more useful in the case of complex problems. In this section, we will discuss some knowledge techniques and methods for knowledge augmentation.

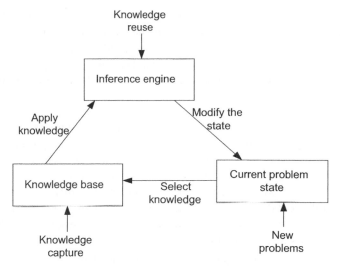

Figure 9.12   Problem-based learning.

9.7.1   Knowledge Augmentation Techniques

Various techniques and methods have been developed to help elicit knowledge from an expert, either a human being or a system. These are referred to as knowledge elicitation or knowledge-augmentation techniques. The techniques that are mandatory for knowledge acquisition can be enhanced and used as knowledge augmentation also. For example, assume that a marketing-business organization contains information about their entire customers in the form of database. This information was gathered with the help of interview techniques [online/offline], by learning from credit card and expense details, by studying carefully the customers buying trend,

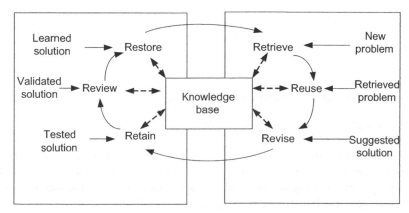

Figure 9.13   Retrieval, reuse, and revisal of knowledge.

and many other communication and information sources. The same techniques can be useful when the new products arrive in the market. By interviewing the same set of customers again related to newly arrived products, the company will gain more knowledge that can be used for learning and strategy forming. Some knowledge-augmentation techniques useful for incremental learning are mentioned below.

- *Protocol-generation* techniques include various types of information gathering and queries (unstructured, semistructured, and structured), reporting techniques (such as self-report and shadowing), and observational techniques. The interviews are conducted by meeting personally the users/clients or customers. Online feedback system/behavioral analysis of customers will also provide very useful client details. If an online feedback system or any other software is used for collecting interview-related data, then the same software will generate various reports and observations.
- *Protocol analysis* techniques are used for knowledge generation based on interviews, knowledge in text form, and similar other forms. This helps to identify knowledge and some important aspects of knowledge such as decision objectives, decision dependencies, relationships, and attributes.
- *Hierarchy-generation* techniques form the hierarchical relationships among different knowledge entities.
- *Sorting* of entities in different groups is essential to create knowledge. This classifies knowledge entities and identifies their relationships.
- *Dependency-based* techniques include the building and use of concept diagrams, state-transition diagrams, event diagrams, and process maps. These diagrams are particularly important in capturing the dependencies and understanding the impact. Conditional probability and similar techniques can be used to identify this dependency.
- *Card sorting* helps us identify similarities between classes of concepts as a part of the model. It is an informal process and works best in small groups.

There are different knowledge-augmentation techniques, and the purpose is to retain the knowledge built in the past and further build knowledge on top of that based on sorting, mapping, and exploration, and with reference to an existing knowledge base.

9.7.2 Knowledge Augmentation Methods

The knowledge augmentation methods are

- *Discover and Share Distinctions and Heuristics*: Apart from the above-mentioned knowledge-augmentation techniques, it is essential to use innovative/inventive approaches for augmentation of knowledge. These methods could be analogy based, mapping based, grouping based, separation practices, and so on. For knowledge sharing, it is essential to promote the exchange of meanings, and then generated knowledge should be reused by all as and when required. Hence naming and relating concepts may be used to promote communication, alignment, and highlight gaps, increase cooperative learning, lift the level of knowledge, and promote the cooperative understanding. Effectively, it aims to improve the overall quality of knowledge base and decision making.

- *Collaborative Knowledge Building and Mapping*: Complete unambiguous details about augmented knowledge needs to be stored in a knowledge base so that knowledge can be made available during new explorations, decision making, and learning. This knowledge will be analyzed and will be reused in the future for understanding meanings, learning, codification, assimilation, and so on.

- *Conducting Cooperative Knowledge Gathering and Inquiry*: Using technology to poll different learners and decision-making algorithms, employing enumerative description techniques to collect information, and sensing environmental intelligence to build the context for decision making are required in cooperative knowledge gathering. Increasing cooperative awareness among intelligent agents and maintaining cognitive diversity to enable multiperspective learning and guarding against perspectives that are not relevant is also considered. This further helps in developing the context in the decision scenario.

- *Write Patterns*: Behavioral pattern captures experience and can help to map proven solutions to common and repetitive problems. This process can apply labels to complex articulations promoting knowledge capture and communication. The sharing of ideas can be made possible through the effective use of patterns. Further patterns also provide abstractions, helping with knowledge transfer giving the holistic and systemic context.

- *Use Patterns*: The group of patterns can have a common link among themselves and can help in spotting connections between different scenarios. This helps in simplifying complex multidimensional problems and identifying learning gaps and finding new solutions through combining and mapping to known patterns. A pattern language bootstraps group agility and intelligence.

9.7.3 Mechanisms for Extracting Knowledge

Two traditional methods, namely induction and deduction, are used to extract knowledge. Deduction (leading away, drawing out) is aimed at particular phenomena, whereas induction (leading or bringing into, introducing) is aimed at the general truth.

9.7.3.1 Deduction   Deduction mechanism is used to prove a particular fact so as to extract knowledge from it. Deduction leads from a generalized statement to a particular fact which helps the user to get more detailed information. The inference mechanisms we discussed in previous chapters can be used for deductions. Deduction is used to test the validity of a statement. It also provides the verification of truthfulness of the statement. An axiomatic-deductive way of thinking is also characteristic of dogmatic-type individuals whose reasoning is based on dogmas postulated either by commonly accepted scientific theories or by scriptures and divine revelations.

9.7.3.2 Induction   An induction mechanism moves from a particular logic to a generalized fact. The induction approach allows for expanding knowledge by adding more facts to it. According to John Stuart Mill, the term "induction" only applies to an inference of an unknown instance or a multitude of instances based on observation of known instances. Inductive methods tend to degrade the validity level of knowledge in the end of each inductive transformation step. Therefore, the validity of initial premises upon inductive transformation in no way warrants the validity of obtained inferences. The works by F. Bacon, J. Herschel, J. S. Mill, and M. I. Karinsky have tremendously pushed forward the development of the system of inductive logic. Contemporary philosopher R. Carnap has made a great input into inductive logic.

There exist various techniques for inductive transformation of knowledge, each with its special features. In this context, it should be important to briefly define the following ones: generalizing induction, inference by analogy (analogy), and cause-and-effect induction. A hypothesis is a special, logical mechanism for obtaining new knowledge. A hypothesis that represents a self-evident assumption is usually called an axiom. For example: to prove that compiler is a system software. The techniques for inductive transformation are discussed below.

1. *Generalizing Induction*: Aristotle wrote: "...induction is a passage from individuals to universals." Generalizing induction is the base of induction as it is based on certain accepted facts. In this example, we generalize the definition of system software and try to induce something from it.
2. *Analogy*: Analogy is the second step of induction and is based on an axiom. An axiom is the statement that is not proved but accepted as truth. An analogy may be either another object with a certain set of similar properties or a specially designed ideal model. The cause-and-effect analysis produces enriched output information, but with a lower validity (truthfulness) level. In this example, we generate an axiom "Compiler is system software."
3. *Cause and Effect*: This step is the actual induction process because it proves the axiom by taking certain known facts and different practical examples. In this example, we use the definition of compiler and system software and prove that our axiom is true and correct.

While building knowledge and systemic learning, we need to use different inductive and inference mechanisms to build the knowledge. The generalization in this case is limited to clusters or group of clusters based on the generalized and representative behavioral pattern. This can be extended to generalization in the case of a decision scenario.

9.8   HEURISTIC LEARNING

Heuristic learning is based on the concept of optimization and hence encounters various problems in it. Data mining and machine learning provide help in the problem of optimization. Data mining is used to extract subjective and nonvolatile content from the previous known facts and empirical details. Data mining stores these contents in a data warehouse. On the other hand, machine learning studies computer algorithms to know more about the problem and devise new techniques to solve them.

Data mining and machine learning primarily identify the problem. Then new methods or techniques are devised to solve them with the help of algorithms. Later, new operators are used for these methods to execute the solution.

According to Daniel Porumbel's paper [8], heuristic learning has the k coloring optimization problem; that is, find a k-coloring that minimizes the number of edges with both ends of the same color (connects).

Data mining or data warehouse techniques with machine learning concepts are basically useful for categorizing information neatly which will be retrieved easily as and when required. For knowledge augmentation to happen successfully, it is essential to implement data mining/data warehouse/algorithms/systems neatly having/possessing advance machine learning behavior.

In addition to these patterns/graphical details based on mined/clustered/categorized data is very useful for analyst to learn from acquired knowledge and augment further.

9.9 SYSTEMIC MACHINE LEARNING AND KNOWLEDGE AUGMENTATION

Machine learning studies computer algorithms for learning while systemic machine learning has the system and its dependencies under consideration while learning. Machine learning is about learning to do better in the future based on what was experienced in the past. But that is not enough and we expect the system to do better in similar, not so similar—complex and not so complex—scenarios. It is based on the observations or data and experience-based learning. The knowledge is augmented with reference to the system and this knowledge is based on actions and impact across the system with reference to different possible decision scenarios.

Knowledge acquisition is integrally tied establishing relationships and more specifically direct relationships among the facts and events. Machine learning considers the basics of human knowledge, and it tries to lead to decisions with reference to direct or mathematical inference. In the case of resolving a particular issue, the mind rebuilds the whole decision scenario, decides learning policies, and make decisions. The most important property of knowledge is that it bears the context and semantically organizes the relevant facts. Learners should be able to build this context and hence should be able to identify these meaningful relationships. These informative inputs along with a decision scenario and systemic parameters should build a decision scenario and overall context for learning and decision making. Machine learning based on the augmented knowledge becomes easier to understand and interpret. Machine learning based on the labeled examples makes it easy for the predicted classification.

9.9.1   Systemic Aspects of Knowledge Augmentation

Knowledge augmentation is about building the knowledge and improving it. The systemic machine learning and systemic knowledge augmentation is required to deal with complex real-life decision making. While learning the parameters, systemic impacts of actions and decisions are considered. In systemic knowledge augmentation, the knowledge is augmented with reference to system and any new knowledge and dependencies are included in knowledge base considering their systemic impact. Systemic knowledge augmentation is depicted in Figure 9.14.

It involves interactive and cooperative knowledge building based on information coming from multiple sources. Systemic knowledge augmentation can be defined as "the impact analysis, processes, tools, and techniques by which the process continuously understands the systemic relationships among different actions and entities." Furthermore, it improves, maintains, and exploits all those elements of its knowledge base and the environmental inputs that are relevant from systemic perspective. It includes the processes, tools, and the infrastructure learnt in systemic context to achieve the goal in decision-space. The knowledge capturing and renewal with reference to decision scenario is depicted in Figure 9.15.

Knowledge augmentation is also an important aspect of intelligent knowledge management. The effective use and mapping of knowledge is required in this process. Furthermore, this faces various strategic and technical issues in the case of strategic knowledge augmentation. These issues include

Knowledge mapping and elicitation

Knowledge transfer and representation

The systemic knowledge transfer and mapping

Internal and external system structures

Knowledge transfer and conversion—different intelligent agents

Knowledge representation in case of different decision scenarios

Maximize the systemic knowledge and systemic value creation

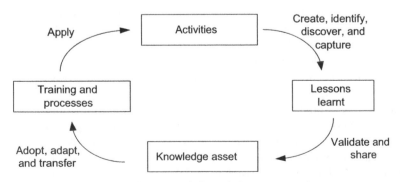

Figure 9.14   Systemic knowledge augmentation.

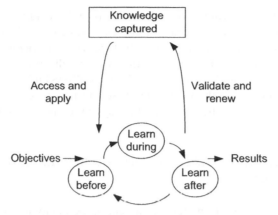

Figure 9.15   Knowledge capturing and renewal.

These different functions and issues related to systemic knowledge augmentation are depicted in Figure 9.16. The major issue related to systemic knowledge augmentation is knowledge conversion with reference to the structure, processes, and event-specific knowledge. Basically the knowledge needs to be checked for reusability and impact.

9.9.2   Systemic Knowledge Management and Advanced Machine Learning

Knowledge management is about making knowledge available and using it effectively. As we discussed in previous chapters and sections, machine learning empowers

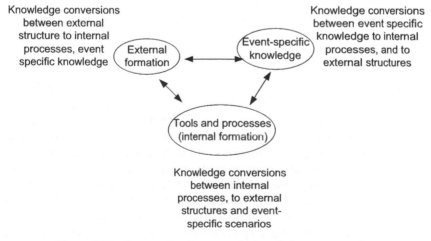

Figure 9.16   Issues related to systemic knowledge augmentation.

knowledge management. Knowledge elicitation is understanding and representing how experts reach a particular decision. Expertise and better decision making many times result based on experience built over the time. This knowledge built by experts over time, along with the ability to use it effectively, results through this experience. Systemic learning as we studied throughout this book allows building knowledge over time and with reference to the knowledge base and new revealed facts. Systemic machine learning proves to be very useful for systemic knowledge management.

9.10 KNOWLEDGE AUGMENTATION IN COMPLEX LEARNING SCENARIOS

The important criterion of implementing and experimenting machine learning algorithms to augment knowledge is to reuse the relevant knowledge after acquisition. Reuse of knowledge will help understand future requirements, effectual decision making, and stressless work environment. The knowledge built in the past is not lost. The results obtained may have entirely different perspectives that can be innovatively implemented across all the numeric domain datasets/applications.

In complex decision problems and real-life scenarios, there is considerable overlap among different decision scenarios. The advent of new information and relationships can change the complete equation, and hence the clusters and mappings built in the past need to be used effectively in light of new information and decision scenario.

9.11 CASE STUDIES

This section explains "how the learning system is built" with computer systems and smart application of advanced machine learning algorithms, patterns, forecasting, and estimating tools and techniques. We have considered three different scenarios from different domains, including finance, software, and sales + marketing as case studies.

9.11.1 Case Study Banking

One bank in the city has very smart, automatic, online software application specially designed for them to cater all needs of their customers. It has a complete setup having high-end servers, individual machines connected in network, huge database system, and so on. When the application started its executions two years ago, initially with the help of a software-development firm and bank employees, the manually stored data were fed in the tables. Once all entries are made up to the mark till date, the system was completely functional, in online and offline mode. Based on this initial information about the bank's customers, loans, business innovative strategies, policies, government norms, and so on, various clusters were formed (with center of cluster, distance between clusters, threshold range, representative series, feature set, etc.), and stored along with a variety of patterns, in database.

While using this application for a variety of purposes on a daily basis, a lot of data/ information will be generated. The data may be related to "opening a new bank account," "transfer of funds," "credit card payment," "loan partial repayment," "open new fixed deposit," "automatically renew fixed deposit," "buy gold," and many more.

Basic clusters need to be updated after collection of new information regarding customers, loans, fixed deposits, gold loans, new accounts, and so on. The freshly acquired knowledge needs to be updated in clusters, so as to analyze customers behavior, for generating new loan schemes, for novel fixed-deposit idea, and so on, by which the bank will increase/expand its products and prove better than its competitors. The bank system will learn about its various products quickly and think about fresh ideas frequently only by studying the generated patterns and due to easy availability of incrementally clustered/augmented knowledge about an entire range of products related to this bank. Figure 9.17 depicts augmentation of knowledge with reference to product building and strategies.

9.11.2 Software Development Firm

Consider an example of a medium-size software development firm. This firm has a set of repeat clients who are completely satisfied with their developed systems, and maintenance is also handled by the same firm. The usual practice followed by this software development firm is to be innovative, give value-added services to its client, maintain their systems, and so on. All details of products/application systems developed by this firm are stored in the form of clusters for effective reuse, and they maintain stress-free development. The acquired knowledge about previously developed software is reused as and when required, to understand the details about upcoming/new project, and to handle new requirements from clients effectively and unambiguously. The project details stored in the form of clusters include cost, development time, team size, resources utilized, domain, category of project (product based/pure software project), links to SRS, other documents/UML diagrams, and so on.

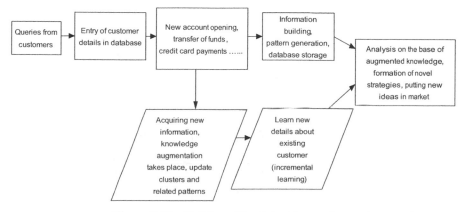

Figure 9.17 Building of bank learning system.

The basic idea behind making use of the cluster approach is easy utilization of augmented knowledge. Incremental learning approach fits best for the software development industry.

The steps involved in software development system learning include

- Gather project-related data.
- Form basic clusters and patterns.
- Accept new requirements from clients (technology change, version change, new functionalities, etc.).
- Update related clusters.
- Reuse augmented knowledge for handling new projects/requirements.
- Apply incremental learning techniques for development of a smart software system.

In next-level application, the cooperative and multiperspective learning is used to collect different perspective and better decision making.

9.11.3 Grocery Bazaar/Retail Bazaar

Grocery bazaar is the chain of grocery stores that contain lots of aisles showcasing various grocery items for easy shopping. They have the best possible databases containing detailed information about all products, repeat client details, customers spending habits, the advantages of variety of deals, the specific time zones and shopping/profits details, shopping points earned by individual customers, redemption of earned points, and other details. This entire information is stored in classified formats, to easily retrieve for decision making, formalizing deals, and so on.

Once again as more and more customers prefer shopping in the grocery bazaar, their details need to be updated in the related database. Expert systems can be made use of for collecting information from clients, who prefer to shop at the grocery bazaar.

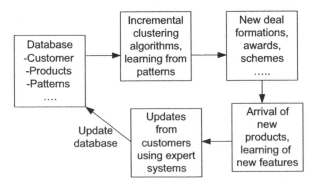

Figure 9.18 Learning steps required for a food bazaar or a grocery store.

To retain the same clients longer, novel ideas need to be generated on regular intervals, keeping the competitors in mind.

Incremental learning and incremental clustering with behavioral patterns in expert systems will be very useful in augmenting and reutilizing knowledge to gain more business.

"How the grocery-bazaar learning system is built" is depicted in Figure 9.18.

9.12 SUMMARY

Knowledge augmentation is one of the basic requirements of effective learning. It is not just the concept of incremental learning. Knowledge augmentation is about understanding the existing knowledge base, decision scenario, and new explored facts. The selective building of knowledge on top of what is learned is knowledge augmentation. The intelligent behavior of the system can be exhibited through effective knowledge augmentation and systemic knowledge augmentation. This chapter described how the knowledge augmentation takes place, along with the systemic aspects of the same. The process of incremental learning starts from the definition of knowledge and its types, and it continues with the process of knowledge acquisition and finally with the incremental process of knowledge augmentation using the knowledge base built in the past.

Continuous data explosion and new technologies struggling to support them is the feature of today's c-world. To handle this exploded data neatly for innovative business approaches and decisions, augmented knowledge will play an important role always, irrespective of the domain to which the organization belongs. The data need to build a context for learning and decision making, while the new data need to be used in the right context for learning. This context building and data mapping in a dynamic environment is all that knowledge augmentation needs. The intelligence of the learning system is the effectiveness of knowledge augmentation process.

REFERENCES

1. Johnson J, Picton P, and Hallam N. Safety-critical neural computing: explanation and verification in knowledge augmented neural networks. Open University, Milton Keynes, 1994 and 2002 [new version], *IEEE Colloquium Safety Critical Neural Computing: Explanation and Verification in Knowledge Augmented Neural Networks*; 1994 and 2002 [new version].
2. Park C, Yu S, and Wang C. Decision making using time-dependent knowledge: Knowledge augmentation using qualitative reasoning. *International Journal of Intelligent Systems in Accounting, Finance and Management*, 2001, **10**(1), 51–66.
3. Lalmas M and Roelleke T. Four-valued knowledge augmentation for representing structured documents. *Lecture Notes in Computer Science*, 2002, **2366/2002**, 237–250.
4. Dazeley R and Kang B. An *Augmentation Hybrid System for Document Classification and Rating. School of Computing, University of Tasmania, Hobart, Tasmania. Lecture Notes in Computer Science*, 2004, **3157/2004**, 985–986.

5. Frommholz I and Fuhr N. Evaluation of relevance and knowledge augmentation in discussion search. University of Duisburg—Essen, Germany. *Lecture Notes in Computer Science*, 2006, **4172/2006**, 279–290.

6. Bodenreider O and Zang S. Knowledge augmentation for aligning ontologies. Semantic Integration Workshop at the Second International Semantic Web Conference, 2003.

7. Chen B, Wang J, and Wang J. Video knowledge augmentation based on summarized contents and online media. *IEEE International Symposium on Circuits and Systems, Taipei, ISCAS*, 2009, 738–741.

8. Porumbel D. *Heuristic Algorithms and Learning Techniques—Applications to the Graph Coloring Problem*, University of Angers, France, Ph.D. Thesis, 2008.

CHAPTER 10

Building a Learning System

10.1 INTRODUCTION

It has been an objective of this book to study the means and ideas to build learning systems that can address some of the problems that are observed by traditional learning techniques. With all the different tools and techniques studied so far, in this chapter we will discuss the process of building the learning system. The objectives of the learning system are to allow use of all information resources and build a framework conducive for learning. There are many systems developed over the years that keep learning from different artifacts. The learning is based on experiences, text information, images, objects, design, words, conversations, and past knowledge. An efficient learning system needs to make use of all this available information effectively. The learning system needs to consider all aspects of data acquisition, machine learning, knowledge building, and knowledge reuse. The concepts of learning from experience, use of exploration, and exploitation of complete knowledge base need to be used while building a learning system.

The learning system building is based on system's best estimate of objective reality. The tasks need to be decomposed to enable learning based on facts that are not observable immediately. Building, sharing, and application of knowledge are fundamental aspects of human intelligence. Any learning and intelligent system should provide these aspects. The drive and willingness to share the information allow the building of knowledge from ancient times. To exhibit this level of intelligence, the system needs to learn form data, relationships, and different system parameters.

10.2 SYSTEMIC LEARNING SYSTEM

A learning system has different components, and these components together with systemic learning algorithms make systemic learning possible. These components include different information sources, knowledge-building components, cooperative

Reinforcement and Systemic Machine Learning for Decision Making, First Edition. Parag Kulkarni.
© 2012 by the Institute of Electrical and Electronics Engineers, Inc.
Published 2012 by John Wiley & Sons, Inc.

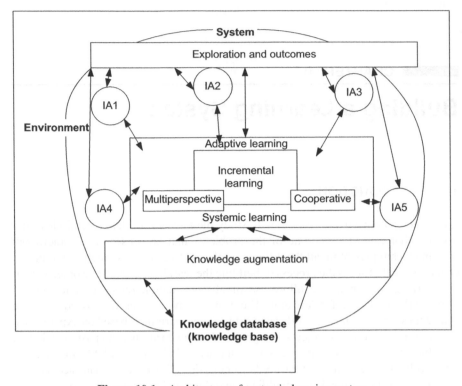

Figure 10.1   Architecture of systemic learning system.

learning, context building, knowledge augmentation, and different other components. The systemic intelligent learning system needs to handle open, dynamic, heterogeneous knowledge building to handle new scenarios effectively. A simple architecture for a systemic learning system is depicted in Figure 10.1.

A set of intelligent agents (IA1 to IA5) interact with the environment, among each other, and with a system along with a systemic learning module to make systemic learning possible. The systemic learning core module has components and algorithms for adaptive learning, incremental learning, and multiperspective learning. To solve problems, humans or any other system require intelligence. This is true even for computers. Learning and knowledge augmentation are the manifestations of intelligence. Intelligence requires dealing with knowledge, along with acquiring and building of relationships. To exhibit this behavior, computers need to acquire knowledge based on the data and information in the environment and system. Machine learning gives this ability to computers. An intelligent system refers to a system capable of acquiring knowledge, containing system information, and integrating this knowledge automatically to deliver decisions. The most important part is learning from experience and exploration. The capability of the system to set itself to go through experience and to learn from new experience is the most important part of it. The intelligence is built through training the system, analytical and behavioral observations, inference, and other means. This training and learning with intelligent knowledge augmentation helps to build a system that can continuously self-improve and thereby exhibit the ability to improve efficiency and effectiveness. The architecture of systemic learning system allows it to closely interact with the environment.

A machine learning system usually starts with some initial training and domain knowledge. This knowledge is acquired by some predefined labeled datasets used for training or human intervention or expert guidance. Furthermore, there is the need to measure both the effectiveness of this knowledge and the accuracy of results produced. The corresponding knowledge organization allows interpreting, analyzing, and testing the knowledge acquired. This can help the learning system to keep track of its learning ability and measure the performance of the system. It is about learning to do better in the future based on what was experienced in the past and building an ability to perform better in even unknown scenarios.

There are different components of learning systems. These include multiagent data acquisition, various learning modules, decision-making modules, sensors, and actuators. Typically, a learning system has the following components:

Learning element (feature analysis, selection, and updating)

Learning policy selection

Knowledge-capture routines

Systemic view and context-building components

Knowledge base

Knowledge acquisition and generation

Knowledge augmentation and reuse

Decision making and relearning

Performance measurement and feedback element

Teacher–trainer—correcting routine based on guidelines

A simple form of learning system with minimum components is depicted in Figure 10.2. This figure includes a learning element and measurement. A feedback system allows for correction in the behavior of the learning system.

Components of the learning system depicted in Figure 10.2 are as follows:

1. *Learning Element—Pattern and Parameters*: It is responsible for learning based on inputs, feedback, and interactions with the environment.
2. *Knowledge Base*: Knowledge base is built based on learning, and the knowledge in the knowledge base is exploited while learning and decision making.
3. *Performance Measurement Element*: Performance element measures the performance based on output.
4. *Feedback Element*: It gives system feedback based on performance and expected outcome of the system.
5. *System to Allow Measurement*: The systemic parameters are measured through interaction with the system.

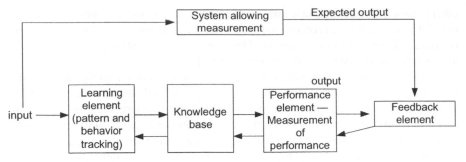

Figure 10.2   Components of the learning system.

Let us discuss them in detail.

10.2.1   Learning Element

The learning element receives and processes the input obtained from an expert or some standard input or from reference material such as magazines, journals, and so on, or from the environment and other systems. This element has different learning algorithms and has an ability to interact with the environment. Furthermore, this element interacts with the knowledge base and uses the knowledge available with it.

10.2.2   Knowledge Base

It contains behavior patterns and historical information. Initially, it may contain some basic knowledge or the domain knowledge available. Thereafter it builds more knowledge based on information received through experience or in the due course of action. While accommodating new knowledge, the existing knowledge is refined. It does not contain just data but basically knowledge built based on that data; the knowledge is represented and stored in the form of relationship, patterns, impacts, possibilities, and clusters.

10.2.3   Performance Element

The performance element tries to measure the performance of the system with reference to expected performance or standard outcome. For any action of exploration in the case of new actions, the performance of the decision needs to be measured. The performance system plays the role of measuring the performance and providing pointers to correct the premise. This element gives feedback for the learning and plays a crucial role in continuous and incremental learning environment.

10.2.4   Feedback Element

Feedback is based on the error. Based on expected performance and actual performance, the feedback is given to improve results and learning. This is the measure to be taken to improve the output and reach close to the expected outcome. This is a typical supervised learning scenario. The feedback is used to determine the corrective action and tune the learning system. There are different feedbacks that any system gets. These feedbacks include feedback from the environment in the case of reinforcement learning, as well as feedback from experts in the case of supervised learning. The feedback in case of reinforcement learning comes in the form of rewards or penalties.

10.2.5   System to Allow Measurement

It is an expert, trained person, or a computer program that is able to produce the correct output and has a proven record to produce the correct results. To check the results produced by a machine learning system, we need to compare the results with some standard or expected outcome for the same input. In the case of more complex systems, there may not be an exact standard system or just a basic standard system that can be used to measure the trend; and even learning allows enhancing the system for more complex learning scenarios.

The learning of the system continues until the expected results are produced. The system in this way is too sensitive to the training set and has high dependence on the expert inputs. The learning and decision-making phases are combined as the system becomes complex and every decision-making phase is used for learning.

There are several factors affecting the performance. They are

- Training set used, and variety in training sets
- Background and domain knowledge of the system and uncertainty
- Feedback mechanism and accuracy
- Dependencies on other parts of system
- Algorithms used and selected

There are other parameters that are not part of the learning system, yet contribute to the performance of the learning system. A few are mentioned below

- Environment and different components
- The techniques to determine dependencies and relationships among different parameters
- The new explorations and prioritization of parameters
- Identifying the decision scenario

This learning can be supervised, semisupervised or unsupervised. Here initially the system learns under supervision and with reference to known scenarios, input data, and environment. This is true in the case of supervised and even semisupervised machine learning. A training set may consist of a carefully chosen variety of training examples for the particular problem or a few known examples from the random set of examples that include a variety of facts and details including mix of relevant data and noise. The proper selection of hypothesis and careful use of background knowledge can optimize the learning by properly selecting and optimizing the search space. The feedback may come in a qualitative or quantitative way. The actionable feedback in the case of learning is preferred. The correct, reliable, and relevant feedback can improve overall learning experience and build knowledge that can provide better decision making. The data or training data comes through sources such as human experts, documents, interactions, and feedback. This comes through reasoning, observations, and behavioral patterns. The environmental/systemic knowledge, content knowledge, and relationships build a context for learning. In a typical learning system there are learners, learning systems, and pedagogy. The scope and each of the components of the system increases as the complex learning and dynamic environment comes into the picture.

Domain, environment, system, and complexity determine the success of machine learning system. Selection of an appropriate algorithm, appropriate training set, and most importantly learning policy can improve machine learning system performance. The learning policy and techniques are selected with reference to the learning objectives and decision scenarios.

10.3   ALGORITHM SELECTION

Algorithm selection refers to choosing the most suitable algorithm for a given goal, among several functionally equivalent algorithms. Proper algorithm selection has to be done for achieving high throughput, low cost, and low-power implementations. The machine learning algorithms that have been extensively applied to text classification (or text categorization) are support vector machines (SVM), k-nearest neighbor (k-NN), Naïve Bayes, Neural Network (Nnet), and Rocchi. Some of these techniques are discussed below.

10.3.1   *k*-Nearest-Neighbor (*k*-NN)

The k-NN algorithm measures the distance between a query scenario and a set of scenarios in the training dataset. The distance between two scenarios is computed using some function $d(x,y)$, where x, y scenarios are composed of n features, such that $X = \{x_1, x_2, \ldots, x_n\}$ and $Y = \{y_1, y_2, \ldots y_n\}$. The training is given by scenarios and data, and these data and scenarios can be represented as multidimensional feature vectors. These training feature vectors are mapped to the expected outcome. These labeled feature vectors are used for training. In the classification phase, unlabeled feature vectors are classified based on nearest training—or labeled sample.

Generally, Euclidean distance is used as the distance metric; other distance measurement methods can also be used based on scenarios and data types. Euclidean distance is calculated as follows:

$$d(x, y) = \sum_{i=1}^{n} \sqrt{x_i^2 - y_i^2}$$

10.3.2   Support Vector Machine (SVM)

The support vector machine (SVM), as a statistical learning theory, has gained popularity in recent years because of its two distinct features. First, SVM is often associated with the physical meaning of the data, so it is easy to interpret. Second, it requires only a small number of training samples. SVM has been successfully used in many applications, such as pattern recognition, multiple regressions, nonlinear model fitting, and fault diagnosis. The essential idea of SVM classification is to (a) transform the input data to a higher-dimensional feature space and (b) find an optimal hyper plane that maximizes the margin between the classes. The group of examples that lie closest to the separating hyperplane is referred to as support vectors.

10.3.3   Centroid Method

Interclass term index and inner-class term index are used to find the centroid. Combination of these indices is used, and a denormalized cosine measure is used to calculate the similarity score between a text vector and a centroid. Given a class Cj of a corpus, there are two classical methods to create Cj's prototype vector. Arithmetical average centroid (AAC) is discussed with the centroid calculation.

$$\overrightarrow{\text{Centroid}_j} = \frac{1}{|C_j|} \sum_{\vec{d} \in C_j} \vec{d}$$

After centroids of different categories are determined, an unlabeled document is classified by finding the closest centroid to the document vector.

$$C' = \text{argmax}_j (\vec{d} \bullet \overrightarrow{\text{Centroid}_j})$$

In centroid-based text categorization, a text in a corpus is represented with a vector space model (VSM), where each text is considered as a vector in term space. A prototype vector (i.e., a centroid) is constructed for each category as a delegate vector for all documents belonging to that class. When classifying an unlabeled document, the vector representing the document is compared with all prototype vectors, and the document is assigned to the class whose prototype vector is most similar.

The performance of centroid-based classifiers depends strongly on the quality of prototype vectors. To improve performance, many studies have attempted using feedbacks to adjust term weight in prototype vectors, such as Drag pushing, Hypothesis Margin, and Centroid. The performance of these adaptive methods is generally better than the traditional centroid-based methods. In particular, some of them can be comparable to SVM classifiers on micro-F1 and macro-F1 evaluations. When considering the domain-specific centroid-based approaches, its results are very effective.

10.4   KNOWLEDGE REPRESENTATION

In the early days of artificial intelligence (AI), people thought that to make a computer intelligent, all one had to do was to give it a capacity for pure reasoning. Scientists soon realized that the exercise of intelligence must involve interaction with an external world, which requires knowledge about that world. The quest for AI inevitably involves the development of methods for endowing computer systems with knowledge. This, in turn, highlighted the question of how to represent knowledge within the computer. Thus the subfield of AI known as knowledge representation (KR) was born. In the AI context, "KR" usually means the quest for explicit symbolic representations of knowledge that are suitable for use by computers. Knowledge is more than just facts, information, or data. These things constitute knowledge only if they are situated in a context provided by some general understanding of the domain they relate to.

Representing knowledge involves representing facts and representing understanding. This generally takes the form of some general model within which the specific facts can be represented and brought into relation with one another. KR is more concerned with formulating such models than with the collection of individual facts, and it is also concerned with establishing a framework of understanding within which the facts make sense.

The key to establishing such a framework is to endow the computer with a capacity for reasoning. KR is really KRR: knowledge representation and reasoning. Armed with knowledge in the form of a collection of general rules and individual facts, a competent reasoner can deduce further individual facts. If we know that Pune is in Maharashtra and that Maharashtra is in India, then we do not need to be told that Pune is in India; we can infer it, provided that we know the rule: For any geographical regions A, B, and C, if A is in B and B is in C, then A is in C; this is also a general rule of inference (Modus Ponens). The formulation of methods of reasoning about knowledge is an important part (or partner?) of KR.

For text categorization, the KR for each document was done initially using the bag-of-words approach.

10.4.1   Practical Scenarios and Case Study

The practical scenarios for text categorization and other classifications can be

- *Optical Character Recognition*: Recognizes images of characters. It also categorizes images of handwritten characters by the letters represented. The handwritten character recognition is also called ICR—intelligent character recognition.
- *Face Detection and Face Authentication*: Finds faces in images (or indicate if a face is present).
- *Spam Filtering*: Identifies email messages as spam or nonspam.
- *Topic Spotting*: Categorizes news articles (say) as to whether they are about politics, sports, entertainment, and so on.

- *Spoken-Language Understanding*: Within the context of a limited domain, determines the context with reference to application and decision scenario.
- The meaning of something uttered by a speaker to the extent that it can be classified into one of a fixed set of categories.
- *Medical Diagnosis*: Diagnoses a patient as a sufferer or nonsufferer of some disease.
- *Customer Segmentation*: Predicts, for instance, which customers will respond to a particular promotion.
- *Fraud Detection*: Identifies credit card transactions (for instance) that may be fraudulent in nature.
- *Weather Prediction*: Predicts, for instance, whether or not it will rain tomorrow.

10.5 DESIGNING A LEARNING SYSTEM

In order to illustrate some of the basic design issues and approaches to machine learning, let us consider designing a text document categorization system that will categorize the text documents based on the predefined categories. Once the system is built, it can be used to determine the category of the future uncategorized text documents. We adopt the obvious performance measure: the percent of text documents the system categorizes into its appropriate category.

The first design choice we face is to choose the type of training experience from which our system will learn. The type of training experience available can have a significant impact on success or failure of the learner. One key attribute is whether the training experience provides direct or indirect feedback regarding the choices made by the performance system.

The supervised machine learning relies on labeled data. This includes initial set $S_o = \{d_1, d_2 \ldots d_s\}$ of documents. These are labeled documents since they were classified previously under specific category $\text{Class}_1 = \{r_1, r_2 \ldots, r_m\}$ with which the system will need to operate. This forms the initial matrix and that is generally correct (Table 10.1).

TABLE 10.1 Training Set and Test Set

	Training Set (Labeled Data)				Test Data			
	D_1	$\ldots$	$\ldots$	d_g	d_{g+1}	$\ldots$	$\ldots$	d_s
R_1	Class_{11}	$\ldots$	$\ldots$	Class_{1g}	$\text{Class}_{1(g+1)}$	$\ldots$	$\ldots$	Class_{1s}
$\ldots$		$\ldots$	$\ldots$	$\ldots$	$\ldots$	$\ldots$	$\ldots$	$\ldots$
r_i	Class_{i1}	$\ldots$	$\ldots$	Class_{ig}	$\text{Class}_{i(g+1)}$	$\ldots$	$\ldots$	Class_{is}
$\ldots$	$\ldots$	$\ldots$	$\ldots$	$\ldots$	$\ldots$	$\ldots$	$\ldots$	$\ldots$
r_m	class_{m1}	$\ldots$	$\ldots$	Class_{mg}	$\text{Class}_{m(g+1)}$	$\ldots$	$\ldots$	Class_{ms}

Note: A training set is a set of documents or labeled samples used for training the system in the case of supervised learning. A test set is used for testing a classifier and testing learning performance. Each document in a test set is tested. It is classified with classifier, and the output is compared with experts' opinion.

10.6 MAKING SYSTEM TO BEHAVE INTELLIGENTLY

The intelligence depends on the knowledge base of the system and the ability and algorithm that allows the system to respond in the case of a new scenario. The act that sounds intelligent at a particular moment may not be a really intelligent act as we realize the drawbacks and side effects of that action over the time. The traditional intelligent system architectures—namely, pattern-based and historical knowledge-based systems—restrict the system performance. Systemic intelligent systems need a different architecture, and they evolve intelligence with exploration.

10.7 EXAMPLE-BASED LEARNING

All supervised learning architectures and algorithms are based on learning with different examples presented in the format that a machine can understand. The assorted examples representing different scenarios are the training set for example-based learning. There can be multiple similar examples with similar outcomes, while in some scenarios the slightly different examples lead to different outcomes. These examples are expected to cover a wide variety of possible scenarios with remarkable decision impact. Examples typically represent decision scenario and the corresponding outcome. There is learning based on worked examples where the steps leading to results are given for learning. Such multiple examples are given for learning. In this case, more and more examples and scenarios are presented for learning. During the new unknown scenario the exploration and outcome is used as an example for learning for the future. The example presented with scenarios and facts helps in building the context. Typically, this method works well in the case of well-formed examples. There is a need to learn beyond examples when the information in examples is not complete. Also, example-based learning helps in building overall context for learning if enough examples are presented for learning with a wide variety. Even multiple learners can learn through their own experience and available parameters. The context is built based on their interaction of different intelligent agents among themselves. Understanding that is developed by each agent through interactions with the environment and other agents along with input in the form of domain knowledge helps to understand the overall system structure. This helps in building the context for decision making. Each agent interprets its own experience rather than solely relying on a knowledge base generated based on past learning. KR, decision making, and reasoning are the three major aspects of this experience-based cooperative learning. Also, there is the need to interpret the knowledge, values, and dependencies.

10.8 HOLISTIC KNOWLEDGE FRAMEWORK AND USE
OF REINFORCEMENT LEARNING

The holistic learning framework interacts with the environment. The purpose as discussed throughout this book is to build decision context. The context here is understanding the situation, different parameters, and dependencies in which learning takes place. This includes representative and relevant parameters and relationships among them. The learners and the intelligent agents actively interact with environment and among themselves. In this interaction they explore the environment, scenarios, and decision parameters. The new experience is interpreted and is used for experience-based learning. The interpretation of new explored facts and experiences by agents is influenced by the previously built knowledge by the learners and agents. Even knowledge built through this interpretation is used to update the knowledge base. This is a continuous process and hence building and augmenting knowledge occurs over time. The holistic learning brings experiences, parameters, context, and decision scenario together to provide a broader picture for decision making. Hence the learning in this scenario is interactive and dynamic. The holistic knowledge framework tries to build knowledge, and in this case learning is essentially interactive knowledge building.

Reinforcement learning essentially tries to bring exploration along with exploitation while learning. Temporal difference learning allows getting frequent rewards and penalties to correct the action for continuous learning. The interactions among the different parts of the system and entities are indicative of contributions of interactions and can help in explaining the parameters and entities in the system in terms of context and weight in decision scenario. Though the context of individual learner is a flexible notion, the cooperative learning notion allows building a context for the decision scenario, which is very important from a learning and knowledge augmentation perspective. To understand interactions that are important from a learning perspective, there is a need to understand environmental properties, system structure, and the decision scenario. The experience and exploration generally comes in the form of actions, and impact of these actions across the system—especially in relationship with decision scenario—is to be considered. In the next phase, acquired knowledge is represented for use and is further applied to the situation for decision making. It is reconstructed and evolved continually with new occasions and activities. Holistic learning interrelates subjects and concepts. It further establishes relationships among these concepts. These concepts are combined to build the overall decision context. Let us first discuss a general machine learning framework and then modify it for holistic learning.

Let us consider a simple system with two subsystems. There are three intelligent agents, a knowledge base, and a knowledge-acquisition module. The environment is sensed with parameters $\{e_1, e_2, \dots, e_n\}$ while the behavior of subsystem Sb_1 is represented by parameters $\{p_{11}, p_{12}, \dots, p_{1m}\}$. Similarly behavior of subsystem Sb_2 is represented by parameters $\{p_{21}, p_{22}, \dots, p_{2m}\}$. A set of behavioral patterns are stored in knowledge base with reference to various decision scenarios along with the recommended actions, forecast, and decisions.

The parameter-selection module tries to select relevant parameters for a decision scenario for each subsystem. Furthermore, all decision parameters are prioritized. For every exploration and new action in an unknown event, the parameters are tracked across the subsystems over the time to calculate rewards and penalties. Every time, based on all information the decision context is determined and that is used to decide the learning policy.

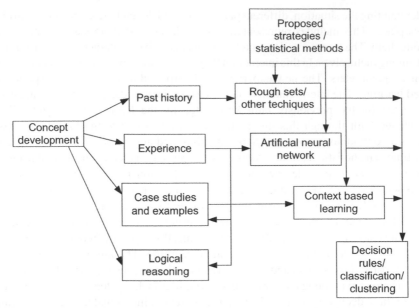

Figure 10.3   Machine learning: general framework and architecture.

The concept can be developed using the different components, say past history or background knowledge, experience, case studies and examples, logical reasoning, and so on. There are different strategies and statistical methods; one example is highlighted in Figure 10.3. They all lead finally to the decision rules, classification, clustering, and so on. They can be used depending on the knowledge to be discovered and the dimension that we see. An appropriate and efficient method is selected — for example, here a rough set can be selected. Generally method is selected based on need to classify or cluster and to arrive at conclusion.

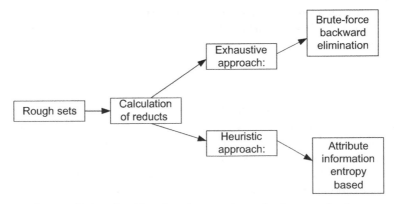

Figure 10.4   Algorithms based on rough sets for feature selection.

Algorithms

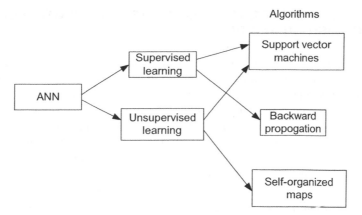

Figure 10.5   Algorithms based on neural networks for seen and unseen data.

10.8.1   Intelligent Algorithms Selection

The algorithms are basically selected based on the strategy or the method you select. For example, if classification is the criteria, it can be done using Rough sets, ANN, Bayesian classification and so on, it depends upon which strategy is chosen. Fuzzy sets can be added for uncertainty, whereas rough sets can be added for imprecision. Figure 10.4 depicts the use of rough sets for feature selection. Figure 10.5 depicts the use of artificial neural networks to handle unseen data.

Logical reasoning and context-based learning can be used for the systemic learning and decision making. Figure 10.6 depicts the use of logical reasoning and context-based learning and different methods used for the same.

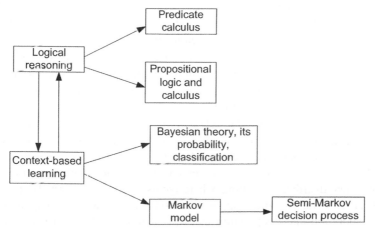

Figure 10.6   Algorithms based on logical reasoning and context-based learning.

10.9 INTELLIGENT AGENTS—DEPLOYMENT AND KNOWLEDGE ACQUISITION AND REUSE

Systemic learning is not possible without efficient acquisition of knowledge and reuse. The efficient acquisition requires intelligent agents. These are typically distributed across the system. These agents sense the environment, outcomes, and actions from different perspectives and have the intelligence to decide local action. Also, these agents need to bring these perspectives where decision making can take place, and they should have the ability to learn cooperatively. Prior to presenting and assessing the individual agent systems and applications domains, let us list the typical agent concepts used in agent technology deployments.

1. *Coordination*: List of agent techniques (based mainly on dedicated coordination protocols and various collaboration enforcement mechanisms) that facilitate coordinated behavior between autonomous, collaborative agents. Coordination usually supports conflict resolution and collision avoidance, resource sharing, plan merging, and various collective kinds of behavior. The different agents coordinate among themselves for knowledge acquisition and prior experience building. The knowledge is shared and enhanced.

2. *Negotiation*: List of various negotiation and auctioning techniques that facilitate an agreement about a joint decision among several self-interested actors or agents. Here we emphasize mainly negotiation protocols and mechanisms and how individual actors shall act and what strategies they shall impose to optimize their individual utility. This is a form of cooperative learning where the negotiation takes place for decision about the optimal strategy.

3. *Simulation*: Techniques that allow inspection of collective behavior of the interactive actors, provided that the models of the individual agents are known. The simulation can be used for learning.

4. *Interoperability*: Agents should interact among themselves. The efficient interoperability helps in cooperative learning and decision making. Agents should be able to work together and understand knowledge built by other agents—they should speak common language to cooperate.

5. *Organization*: Techniques that support agents in their ability to organize autonomously in permanent or temporal interaction and collaboration structures (virtual organizations), assign roles, establish and follow norms, or comply with electronic institutions.

6. *Meta-reasoning and Distributed Learning*: In the multiagent community, there are various methods allowing an agent to form a hypothesis about available agents. These methods work mainly with the logs of communication or past behavior of agents. The agent community also provides techniques for collaborative and distributed learning, where agents may share a learned hypothesis or observed data. A typical application domain is distributed diagnostics. The collaboration and cooperation takes place through different communication means.

7. *Distributed Planning*: Specific methods of collaboration and sharing information while planning operation among autonomous collaborating agents. The agent community provides methods for knowledge sharing, negotiation, and collaboration during the five phases of distributed planning—task decomposition, resource allocation, conflict resolution, individual planning, and plan merging. These methods are particularly suitable for the situations when the knowledge needed for planning is not available centrally. The information is shared through common datasets or some other means.

8. *Knowledge Sharing*: Techniques assist in sharing knowledge and understanding different types of knowledge among collaborative parties as well as methods allowing partial knowledge sharing in semitrusted agent communities (closely linked with distributed learning and distributed planning).

9. *Trust and Reputation*: Methods allow each agent to build a trust model and share reputation information about agents. Trust and reputation is used in a noncollaborative scenario where agents may perform nontrusted and deceptive behavior.

For "systemic learning," there is a need to deploy intelligent agents and identify all information sources. Further intelligent agents need to possess the above properties for systemic knowledge acquisition and cooperative decision making.

10.10   CASE-BASED LEARNING: HUMAN EMOTION-DETECTION SYSTEM

We considered earlier that machines are trained like humans to interpret human emotions. There can be various aspects as to how the machines are trained to achieve this. All the strategies discussed above can be implemented, and the best suitable and relatively better results strategy is chosen. A generalized case-based learning architecture is depicted in Figure 10.7.

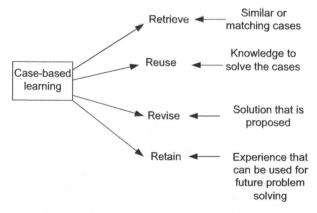

Figure 10.7   Case-based learning architecture.

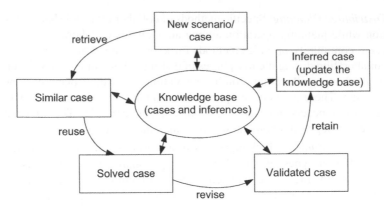

Figure 10.8 Case-based learning cycle.

The case-based learning has a knowledge base full of cases and inferences. The previous case or the case in knowledge base is used as a reference. The knowledge base is used and enhanced as further learning takes place. In the case of a new scenario, the similar case is retrieved from the knowledge base. The cases are classified with reference to relevance and similarity. Furthermore, the learning takes place based on new cases, similar cases, and not so similar cases with reference to exploration, experience, and outcome. A typical case-based learning cycle is depicted in Figure 10.8.

When we take an example of an emotion-detection system, it can use various methods and may have different components. A simple feature-based approach with a traditional bag of methods is depicted in Figure 10.9. The same problem can be handled using a systemic and holistic approach where all the information across the system and subsystems can be used for learning. The impact of any action across the subsystems and inference in a holistic way can help to improve the accuracy of detection. Different methods for feature classification and decision making for emotion detection are depicted in Figure 10.9.

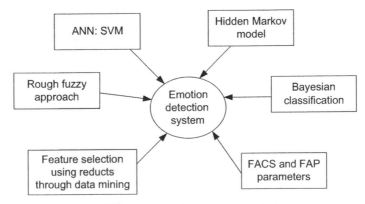

Figure 10.9 Different strategies for emotion-detection system.

10.11 HOLISTIC VIEW IN COMPLEX DECISION PROBLEM

Let us consider the same case study as discussed above.

- The decision is formulated taking into consideration the different parameters that affect the system. The various parameters may help us to formulate a better machine learning system, especially if the system is complex.
- For example, there are various parameters related to emotion that can help in building the systemic view for emotion detection.

These parameters include

1. *ECG (electrocardiogram)*: Output and the patterns during the period of observation. This includes the following observations:

   Heart rate (HR), interbeat intervals (IBI), heart rate variability (HRV), and respiratory sinus arrhythmia.
   Emotional cues
   - Decreasing HR: relaxation, happy
   - Increasing HRV: stress, frustration

   There can be more cues formed based on patterns and explorations.

2. *BVP (blood volume pulse)*: This includes photoplethysmography that bounces infrared light against a skin surface and measures the amount of reflected light and Palmar surface of fingertip.

   The observations or the features are: HR, vascular dilation (pinch), vaso‑constriction.
   Emotional cues
   - Increasing BV—angry, stress
   - Decreasing BV—sadness, relaxation

3. *RESP (respiration)*: This includes a relative measure of chest expansion, on the chest or abdomen, respiration rate (RF), and relative breath amplitude (RA).
   Emotional cues
   - Increasing RF—anger, joy
   - Decreasing RF—relaxation, bliss

   Based on all these parameters, multiple combinations are possible and the emotional context and systemic learning can be made possible.

4. *Temp (peripheral temperature)*: This includes
   - Measure of skin temperature as its extremities
   - Dorsal or palmer side of any finger or toe
   - Dependent on the state of sympathetic arousal
   - Increase of Temp: Temperature is higher in the case of anger as compared to happiness. Similarly, temperature is higher in the case of sadness as compared to surprise or disgust.

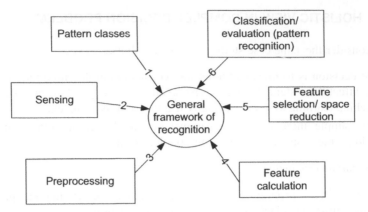

Figure 10.10   Holistic view for decision making for the emotion-recognition system.

With all these parameters the overall systemic knowledge is built. This knowledge allows inferring systemic dependencies with reference to decision scenario. The decision scenario may have the environment under observation. This may have scenarios such as

Person about to deliver speech
Person working in office
Person facing interview
Person getting ready for a match
Person receiving the cup for winning major tournament
Person in discussion with someone

The above parameters along with environment build a context for decision making. The learning based on decision scenario and systemic parameters help in producing better results. Figure 10.10 depicts the general framework for the emotion-detection system.

As shown in Figure 10.10, the functions of each module are as follows:

Pattern Classes: Performs supervised classification.

Sensing: Responsible for data acquisition using biosensors in natural or scenarized situation.

Preprocessing: Performs noise filtering, normalization, up/down sampling, segmentation.

Feature Calculation: Extracts all possible attributes that represent the sensed raw bio signal.

Feature Selection/Space Reduction: Identifies the features that contribute more in the clustering or classification.

Classification/Evaluation (Pattern Recognition): Concerned with multiclass classification.

10.12   KNOWLEDGE REPRESENTATION AND DATA DISCOVERY

Knowledge data discovery, KDD, is basically extraction of knowledge from data, through information, and towards building and exhibition of intelligence. Machine learning is a part of AI, which in simple words is the intelligence related to machines, computers being more specific. The knowledge discovery process is a sequence of steps that help in discovering knowledge from available data and knowledge from different information and knowledge sources. Understanding data and problem domain are very important in the knowledge-discovery process. Since computers can be made to understand relationships through mathematics, they have to be made intelligent artificially through models based on contextual relationships and mathematics. The KR is one of the most important aspects of it. The knowledge should be represented in usable form and should be used effectively by the learning components.

A machine does not have emotions, but learning fed to the system can even generate that through a machine learning system. In human beings, different emotions can be assumed as different emotional perspectives and knowledge sensing with reference to environment and decision scenarios. The effective data mining, along with distributed and cooperative learning, can help in determining these emotions. Holistic knowledge discovery models integrate the whole process with metadata, context, and different knowledge sources with reference to different decision scenarios.

Data mining, which has a wide scope of applications, uses algorithms for extracting information from a large set of data. The same information to be retrieved in better form, known as knowledge and then intelligence through a modified version of data mining algorithms on a small set of data, is created in machine learning.

Machine-learning research has provided opportunities for building intelligent products in different domains. There are various open problems such as (1) applying a number of classifiers and selection of appropriate learning policy with reference to applications, (2) exploration-based reinforcement learning in practical scenarios, (3) applying supervised learning for high-dimensional dynamic learning scenario, and (4) development and use of complex stochastic models.

Machine learning is basically a field of science for programming systems in order to automate them through learning and training that may be with experience, with number of samples, with time, and so on, as a human does. But due to its high computation power, the results may be extremely high and unexpected, getting through the drawback of humans. In fact, a multiperspective approach that itself can improve and show magical results may not be possible with humans due to their limit of capacity in certain areas and can be overcome through machines. The mobile robots, intelligent networks, and intelligent traffic control are some examples where the capabilities of machine learning can be used, and with systemic learning these applications can handle more complex scenarios and learning problems. "Mobile robots" navigate based on the training and experience provided earlier and try to capture more with the help of sensors around the room to get the best appropriate results. Basic algorithms are selected based on this application, which include general conjectures, grasp-based techniques, selection criteria, and comparison based on learning and experience criteria and are then designed.

The majority of work to date in machine learning has focused on learning from (a) examples represented as attribute vectors, where each attribute is a single number or symbol, and (b) a single table that contains all the vectors. However, much (or most) of the data in KDD applications is not of this type. For example, relational databases typically contain many different relations/tables, and performing a global join to reduce them to one without losing information is seldom computationally feasible. (Inductive logic can handle data in multiple relations, but simultaneously focuses on learning concepts that are themselves in first-order form, thus addressing a doubly difficult problem.) The World Wide Web is mostly composed of a combination of text and HTML, plus image and audio files. The data recorded by many sensors and processes, from telescopes and Earth-sensing satellites to medical and business records, have spatial and temporal structure. With regard to customer behavior, as well as mining applications that are of central concern to many companies, people can be hierarchically aggregated by occupation and other characteristics, products by category, and so on. Simply converting data of all these types to attribute vectors before learning, as is common today, risks missing some of the most significant patterns. Although in each case, traditional techniques for handling these types of data exist, they are typically quite limited in power compared to the machine-learning algorithms available for the attribute-vector case. There is much scope for extending the ideas and techniques of machine learning in this direction.

A machine learning system appropriate to future KDD applications should be able to function continuously, learning from an open-ended stream of data and constantly adjusting its behavior while remaining reliable and requiring a minimum of human supervision. The future is likely to see an increasing number of applications of this type, as opposed to the one-shot, stand-alone applications common today. Early indicators of this trend are (a) e-commerce sites that potentially respond to each new user differently as they learn his/her preferences and (b) systems for automated trading in the stock market. The trend is also apparent in the increasing preoccupation among corporations to instantly and continuously adapt to changing market conditions, leveraging for this purpose their distributed data-gathering capabilities. While there has been some relevant research in machine learning, learners of this type must address several interesting new issues. One is smoothly incorporating new relevant data sources as they come online, coping with changes in them, and decoupling from them if they become unavailable. Another is maintaining a clear distinction between two types of change in the learner's evolving model(s): some that are simply the result of accumulating data and consequently progressing in the learning curve, and others that are the result of changes in the environment being modeled.

In KDD applications, learning is seldom an isolated process. More typically, it must be embedded into a larger system. Addressing the multiple problems, this raises an opportunity for machine learning to expand its focus and its reach. The need to efficiently integrate learning algorithms with the underlying database system creates a new interface between machine learning and database research such as (a) finding query classes that can be executed efficiently while providing information useful for learning and (b) simultaneously finding learning approaches that use only efficiently executable queries. Some relevant questions are: What types of sampling can be efficiently supported, and how can they be used? What is the best use that can be made of a single sequential scan of the entire database? The outcome of this iterative process may be query types and learning algorithms that are both different from those known today. The interface between machine learning and databases also involves the use for learning purposes of the metadata that is sometimes available in database systems. For example, definitions of fields and constraints between their values may be a valuable source of background knowledge for use in the learning process.

To be used to its full potential, KDD requires a well-integrated data warehouse. Assembling the latter is a complex and time-consuming process, but machine learning can itself be used to partially automate it. For example, one of the main problems is identifying the correspondences between fields in different but related databases. This problem can be formulated in learning terms. Given a target schema $\{X_1, X_2 \ldots X_n\}$ and examples of data in this schema, induce general rules as to what constitutes an X_i column. Given a table in a source schema $\{Y_1, Y_2, \ldots Y_n\}$, the goal is now to classify each of the Y columns as one of the X's (or none), with the results for one Y potentially constraining those for the others. Data cleaning is another key aspect of building a data warehouse that offers many research opportunities for machine learning.

Very large databases almost invariably contain large quantities of noise and missing fields. More significantly, noise is often of multiple types, and its occurrence varies systematically from one part of the database to another (e.g., because the data comes from multiple sources). Similarly, the causes of missing information can be multiple and can vary systematically within the database. Research-enabling machine learning algorithms to deal with noise and missing data was one of the main drivers of their jump from the laboratory to widespread real-world application. However, example-independent noise and missing data are typically assumed. Modeling systematic sources of error and missing information, and finding ways of minimizing their impact, is the next logical step.

The need to produce learning results that contribute to a larger scientific or business goal leads to the research problem of (a) finding ways to integrate these goals more deeply into the learning process and (b) increasing the communication bandwidth between the learning process and its clients beyond simply providing class predictions for new examples. The importance in KDD of interaction with the human user (expert or not) gives a new urgency to traditional machine learning concerns such as comprehensibility and incorporation of background knowledge. Today's multiple KDD application domains provide a wealth of driving problems and testing grounds for new developments in this direction. Many major application domains (e.g., molecular biology, Earth sensing, finance, marketing, fraud detection) have unique concerns and characteristics, and developing machine learning algorithms specifically for each of them is likely to occupy an increasing number of researchers.

Most machine learning research to date has dealt with the well-circumscribed problem of finding a classification models. These models typically are given a single, small, relatively clean dataset in attribute-vector form. These attributes are defined in advance and chosen to facilitate objective-based learning. In these cases the end-goal (accurate classification) is simple and well-defined.

10.13 COMPONENTS

Machine learning is more of concept learning which depends on the application on which it has to be developed. There are no standard algorithms or components designed for all applications. But then, in general, experience says that the system is trained accordingly. So first, the concept learning and its components have to be learned, and then how each component has to be trained needs to be learned. For that different statistical algorithms are selected or may be designed and modified depending on the requirements. A simple example of systemic learning in case of design of architecture is discussed below.

10.13.1 Example

Interactive design of a physical (construction/building or similar) system is depicted in Figure 10.11. Here the system has various components including users, environment, and different subsystems. Design setting, aesthetic references, load it needs to bear, future expansion, and impact on other systems such as agricultural zones, ecological systems, traffic scenarios, and budgetary provisions should provide rewards for every explorations. The interactive design of a physical system with the aid of systemic intelligence and learning is shown in the figure. It gets inputs from different subsystems. Further optimization and monitoring allows to keep a continuous track of development.

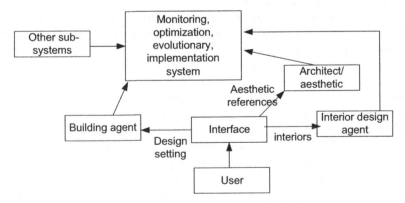

Figure 10.11   Interactive design of a physical system.

10.14   FUTURE OF LEARNING SYSTEMS AND INTELLIGENT SYSTEMS

The development of complex systems and integrated and multidisciplinary applications demand systemic, incremental, and multiperspective machine learning. The true intelligent security systems are not confined to signatures or data available at hand but instead overall system behavior. The applications of these learning systems include integrated security systems, integrated education system, intelligent business-decision systems, and so on. The capability to adapt to new environment and changes is going to be the key in these systems. The learning system should evolve and build better abilities with explorations. More and more information that is available at disposal needs to be used to build knowledge for decision-scenario-specific learning as well as decision making. The true intelligent system adapting to unknown and new scenarios that can learn from limited information and can resolve complex decision problems with multiperspective and cooperative learning is the future of intelligent systems.

The paradigm of the future learning systems is not based on historical information and signatures only. This paradigm really limits the usability and intelligence of machine learning systems. The development and design of new learning systems should not only handle large information at hand gracefully, but should learn from experience and beyond experience. Whole-system learning can offer the required platform for the next-generation systems where all the information, inferences, and adaptation available at disposal are used for learning in proper context. Apart from that, next-generation intelligent systems would need better knowledge-acquisition mechanisms to collect knowledge and data—knowledge building. Systemic machine learning is about learning beyond event, inferring beyond the data, and evolving beyond the immediate response. High complexities and interdependencies are the challenges in front of the system. The new paradigms and knowledge-centric systems can empower the learning system to meet these challenges to offer next-generation intelligent and learning systems that can learn from experience, use knowledge effectively, understand interdependencies, and truly help in building an intelligent system that is not restricted by visual time and space boundaries.

10.15   SUMMARY

This chapter presented the concepts and architectures to build intelligent and learning systems based on paradigms discussed and introduced in this book. Intelligent systems are focused on knowledge building, knowledge acquisition, and exploration-based learning. Learning is a continuous process and is confined only by decision scenarios and availability of data. Going beyond data and exploring beyond view along with inferring on timescale and space is required for systemic machine learning. The different learning methods, multiperspective learning, and adaptive learning to decide learning policy makes the learning in dynamic scenarios possible. Cooperation among different learning components and interactive learning is discussed in this chapter. Understanding learning and developing learning policy has been a challenge in front of researchers, whether it is educational systems or any other complex application. Machine learning includes statistics, psychology, computing, and system structures. With high complexity and huge research opportunities in these areas, it has created a real difficulty in selection of methods but opened avenues for integrated research. The integrated approaches and systemic learning can help to build foundation to build next-generation intelligent system. In systemic machine learning, complexity and data will create unbounded learning opportunities rather than restricting learning by events and uncertainties.

Statistical Learning Methods

Statistical learning methods are used for problems wherein we are uncertain about the outcome. Hence probability-based methods fall under this category. In this section, we will discuss the statistical learning methods such as the Bayesian Classification with the basics of probability.

A.1 PROBABILITY

Let us start with the probability. A sample space is defined as a set of all possible outcomes. An event is defined to be a subset of the sample space. Consider p as the probability of outcome. Then the probability of any event "x" which is a subset of sample space S is defined as

$$P(x) = \frac{n(x)}{n(S)}$$

where $n(x)$ is the number of elements in x and $n(S)$ is the number of elements in the sample space.

Broadly speaking, the types of events that can occur are classified into two categories: (1) mutually exclusive and (2) independent.

A.1.1 Mutually Exclusive Events

Events are said to be mutually exclusive if they cannot occur simultaneously. They can be said to be dependent events. The sum of mutually exclusive probabilities is always 1. Let us consider events x and y.

If x and y are the events that are mutually exclusive, then

$$P(x \, or \, y) = P(x) + P(y)$$

Reinforcement and Systemic Machine Learning for Decision Making, First Edition. Parag Kulkarni.
© 2012 by the Institute of Electrical and Electronics Engineers, Inc.
Published 2012 by John Wiley & Sons, Inc.

A.1.2 Independent Events

Events are said to be independent if one event is not related or has no influence/does not affect the outcome of another event. Consider the events x and y. If x and y are independent events, then the probabilities will be defined as

$$P(x \text{ and } y) = P(x)P(y)$$

Note: Consider the events x and $\sim x$ which are mutually exclusive. They are not independent events. So, if the event x has taken place, there is no question of $\sim x$ taking place.

A.1.2.1 Conditional Probability   Conditional probability forms the basis for statistical learning methods. Conditional probability is defined as probability of event x, given event y which has already occurred. It is represented as $P(x|y)$ (probability of x, given y).

For events x and y that are not independent, where y is given/has occurred already, then the probability is calculated as

$$P(x|y) = \frac{P(x \text{ and } y)}{P(y)}. \tag{A.1}$$

The corollary is

$$P(x \text{ and } y) = P(x|y)P(y)$$

For events that are independent, the conditional probability for x, given y, is defined as

$$P(x|y) = \frac{P(y)P(x)}{P(y)}$$

Hence $P(x|y) = P(x)$.
Some probabilities formulas: For events x and y

(1) The product rule is

$$P(x \text{ and } y) = P(x|y)P(y) = P(y|x)P(x)$$

(2) The sum rule is

$$P(x \vee y) = P(x) + P(y) - P(x \text{ and } y)$$

A.2 BAYESIAN CLASSIFICATION

Let us move towards discussion of Bayesian classification. Bayesian classification falls under the statistical classification that has a probabilistic approach. It predicts membership of class, depending on the probabilities. Bayesian classification is based on Bayes' theorem, discussed below. Bayes' theorem or Bayes' rule is named after Thomas Bayes. The theorem puts forth the conditional probability most often called "posterior probability." This is calculated on the basis of the prior probability. The problem is typically to determine the best hypothesis, given some training data.

Consider $P(h)$ to be the initial probability (h is some hypothesis). This is before training data are made available. This is often called as prior probability of h or also referred as marginal probability of h.

$P(x)$ to be prior probability for the training data—x. Here no knowledge about the hypothesis is available. It is also referred to as the marginal probability of x.

Now, $P(x|h)$ will be the probability after observing the training data with some given hypothesis.

The calculation of the posterior hypothesis will be done as follows:

$P(h|x)$: probability that h holds, given some observed training data; this will be calculated as

$$P(h|x) = \frac{P(x|h)P(h)}{P(x)}$$

This is Bayes' Theorem. Bayes' theorem exhibits the relationship between conditional probabilities.

Bayesian learning helps in adding up the predictions to the existing knowledge base which will help in further classification of new data.

Derivation of Bayes' Theorem (on the basis of conditional probability)

Let us start with the conditional priority for the events x and y.

From Equation (A.1), we know that probability of x, given y, is

$$P(x|y) = \frac{P(x \text{ and } y)}{P(y)}$$

Likewise, probability of y, given x, will be expressed as

$$P(y|x) = \frac{P(x \text{ and } y)}{P(x)}$$

From the above two equations, we have

$$P(x \text{ and } y) = P(x|y)P(y) = P(y|x)P(x)$$

Hence we get the Bayes' theorem

$$P(x|y) = \frac{P(y|x)P(x)}{P(y)}$$

A.2.1 Naïve Bayesian Classification

The Naïve Bayesian classifier works on Bayes' theorem. In the Naïve classification, the variables are treated to be independent. It considers that all the properties are not related to each other and hence the classification is independent in a probabilistic way. In supervised learning methods, Naïve Bayes is considered to be a potential method for the classification.

Let us understand the working of Naïve Bayes:

Consider "T" to be training set; with the labeled classes. The training set constitutes of tuples say D with attributes vector d_1 to d_n; represented as

$$D = \{d_1, d_2, \ldots, d_n\}$$

where the attributes are $Ab = \{Ab_1, Ab_2, \ldots, Ab_n\}$

Assume that the classes available are C; from C_1 to $C_{\max}$.

$$C = \{C_1, C_2, \ldots, C_{\max}\}$$

Now, given a new data "N," the job of the classifier is to predict the class for it. It can be represented as $N = \{nd_1, nd_2, \cdots, nd_n\}$.

The Naïve Bayes predicts class on the basis of "highest posterior probability." Let us say that the class predicted is C_i. This is done with the following rule:

$$P(C_i|N) > P(C_j|N)$$

such that $j \neq i$ and $1 \leq j \leq$ max (the total number of classes).

So, we have to maximize $P(C_i|N)$. This class C_i is at times referred to as "maximum posteriori hypothesis."

With Bayes, we have

$$P(C_i|N) = \frac{P(N|C_i)P(C_i)}{P(N)}$$

Since $P(N)$ is constant, as it does not depend on C, we need to concentrate only on the numerator. So we can infer that when the numerator value is maximized, so is the outcome.

Assuming that the marginal (prior) probabilities of class are not available, we can say that $P(C_1) = P(C_2) = \cdots = P(C_m)$.

Considering the above-mentioned two conditions of the denominator $P(N)$ and the classes being constant, we further infer that we have to maximize $P(N|C_i)$.

If we have sets of data with a considerable number of attributes, then in terms of computation it is an issue to be looked upon. Calculation of $P(N|C_i)$ will be a costly venture. Here comes into the picture the independent assumption of Naïve. Considering this, we have

$$P(N|C_i) = \prod_{p=1}^{n} P(nd_p|C_i)$$

$$= P(nd_1|C_i)P(nd_2|C_i)\ldots P(nd_n|C_i)$$

Computing the values of $P(nd_1|C_i)$ onwards can be done from the training set available. We have already stated that the attributes $nd_1, nd_2, \ldots$ onwards are the actual values for the attribute. It is necessary to decide the type of category of the attribute. The attribute can be

(1) categorical or
(2) continuous.

For the calculation of $P(N|C_i)$, two cases are to be considered:

Categorically, $P(nd_p|C_i)$ refers to the number of tuples of class C_i, divided by $|C_i, T|$, which refers to the total number of tuples/sets in class C_i.

In the case of continuous valued, the Gaussian distribution needs to be considered. The attribute here is considered to have Gaussian distribution with deviation ∂ and mean μ:

$$g(nd, \mu, \partial) = \frac{1}{\sqrt{2\pi\partial}} e^{-(nd-\mu)^2/2\partial^2}$$

Then we have

$$P(nd_p|C_i) = g(nd_p, \mu C_i, \partial C_i)$$

The posterior can be written as

$$P(C_i|N) = P(C_i) \prod_{p=1}^{n} P(nd_p|C_i)$$

So with the above rule, we label the new data N to class C_i that will have highest posterior probability. To determine the class of N, we need to evaluate $P(N|C_i) P(C_i)$ for each class C_i. The label is predicted if

$$P(N|C_i)P(C_i) > P(N|C_j)P(C_j)$$

where j ranges from 1 to max and is not equal to i.

A.2.2 Pros and Cons of Bayesian Classifiers

It is found that Bayesian classifiers are comparable with the decision trees and *NN* classifiers considering some domains. But there are some drawbacks wherein the dependency is on the available probability data. At the same time, the independence of attributes considered also results in low accuracy rate. Still, Bayesian methods put forth justification steps to support the outcomes.

A.3 REGRESSION

Numeric prediction is often referred to as regression. Numeric prediction is the prediction of the numeric data, whether it is continuous or categorical. Regression analysis models the relationship between two types of variables that can be independent and dependent. The independent variables are referred to as predictor variables, and the dependent ones are response variables. The predictor variables are the attribute vectors and whose values are available well in advance. Amongst the different regression techniques available, linear is widely used. Let us discuss these techniques.

A.3.1 Linear

This has a response variable: y and a predictor variable x and is represented as

$$y = a + bx$$

where a and b are regression coefficients. They can also be mapped as values or weights, represented as

$$y = v_0 + v_1 x$$

Consider T is the training set comprising predictor variables $x_1, x_2, \ldots$ and $y_1, y_2, \ldots$. The training set has pairs such as $(x_1, y_1), (x_2, y_2) \ldots x|T|, y|T|$. Calculation of the regression coefficients is done with x and y as the means of predictor and response variables, respectively.

$$v_1 = \frac{\sum_{i=1}^{|T|}(x_i - \overline{x})(y_i - \overline{y})}{\sum_{i=1}^{|T|}(x_i - \overline{x})^2}$$

$$v_0 = \overline{y} - v_1 \overline{x}$$

A.3.2 Nonlinear

When the relationship between the predictor and response variable can be represented in terms of a polynomial function, we use nonlinear regression. It is also referred to as polynomial regression. We use polynomial regression when there is just one predictor variable, where the terms of polynomial are added to the linear ones. With the transformation methods we can convert the nonlinear ones to linear ones.

A.3.3 Other Methods Based on Regression

We have generalized models that characterize the basis on which linear regression can be applied to categorical variables. The response variable y here is a function of mean values for y. There are different types of generalized models; the most commonly used are:

(1) Logistic—Here the probability of some event that occurs as a part of linear function comprising of set of predictors is considered.
(2) Poisson—It seeks to model count, typically the log of the count. The probability distribution is different here than the logistic one.

We also have log-linear models that are used in natural language processing. They assign joint probabilities to the observed datasets. In the log-linear method, all attributes are required to be categorical. It is also used in data compression techniques.

The other approach is decision tree induction. This method is suited for the prediction of continuous-valued data. The types are regression and model trees. The leaf node contains the continuous-valued prediction, whereas, in model trees, each leaf node constitutes regression model. It is found that regression and model trees exhibit more accuracy than linear regression.

A.4   ROUGH SETS

Rough sets are used as basic framework for areas of soft computing. It is oriented towards approximation to get the low-cost solutions. This happens in cases where exact data are not required. So rough sets are used to get solutions in areas where the data are noisy, the data types do not belong to a particular type but instead are a mixture of different variants, the data are not fully available, or the data are huge and there is need to use the background knowledge.

Rough sets provide mathematical tools that are used to discover the hidden patterns. Since they try to identify or recognize the hidden patterns, they are typically used in feature selection and extraction-based methods. We can say that they aim at "knowledge discovery." They are gaining more and more importance in data mining, with a specific lookout towards multiagent systems.

Pawlak [1,2] introduced the rough sets to represent the knowledge and to find out the relations between the data. In information systems, we have classes of objects where it is not possible to distinguish the terms available. They need to be roughly defined. Rough set theory is based on equivalence relations. These relations partition the data into equivalence classes and comprise an approximated set with lower and upper bounds. Let us consider the information system representation.

$$IS = \langle U, A, V, f \rangle$$

where U is the nonempty finite set of objects, represented as

$$U = \{x_1, x_2, \ldots, x_n\}$$

A is a nonempty finite set of attributes, where V_a is the value of the attribute a.

$$V = U_{a \in A} V_a$$

f is a decision function such that $f(x, a) \in V_a$ for every a that is an element of A and x an element of U.

$$f : U \times A \to V$$

A.4.1   Indiscernibility Relation

Let us move towards the discussion of an equivalence relation. A binary relation R is said to be equivalence if it is reflexive, symmetric, and transitive.

So, $R \subseteq X \times X$.

xRx for any object in x is satisfied. If xRy, then yRx holds; and if xRy and yRz, then xRz also holds. The equivalence class $[x]_R$ of element x belonging to X consists of objects y belonging to X such that xRy.

Let IS be the information system, then with any B which is subset of A, there is equivalence relation it is associated with represented as

$$\text{IND}_{\text{IS}}(B) = \left\{ (x, x') \in U^2 | \forall a \in B, a(x) = a(x') \right\}$$

If the elements $(x, x') \in \text{IND}_{\text{IS}}(B)$, then x and x' are said to be indiscernible from each other. B is the indiscernibility relation, and its equivalence classes are represented as $[x]_B$.

With the equivalence classes, U is split into partitions, which can be used to generate new sets.

A.4.2   Set Approximation

Consider IS as the information system with B as subset of A and X subset of U. We can approximate X with the use of information in B by generating the upper and lower bounds or approximations. The upper and lower approximations here are B-lower and B-upper represented as $\underline{B}X$ and $\overline{B}X$, where

$$\underline{B}X = \{x | [x]_B \subseteq X\} \text{ and}$$

$$\overline{B}X = \{x | [x]_B \cap X \neq \phi\}$$

A.4.3   Boundary Regions

The boundary region for X is defined as

$$\overline{B}X - \underline{B}X$$

$U - \underline{B}X$ is the negative region and $\underline{B}X$ is the positive region represented as POS_B.

A.4.4   Rough and Crispy

A set is said to be rough if its boundary region is not empty; otherwise it is said to be a crisp set.

A.4.5   Reducts

We are concerned about the attributes that preserve the indiscernibility and hence the approximations. There can be many such attribute sets or rather subsets. The subsets which are minimal are called reducts.

A.4.6   Dispensable and Indispensable Attributes

The attribute a is said to be an indispensable attribute if

$$\text{IND}(A) = \text{IND}(A - \{a\})$$

Otherwise it is said to be indispensable.

If removal of an attribute results in inconsistency, then that attribute is used as a CORE. It is represented as

$$\text{CORE}_B(A) = \{a \in A : \text{POS}_A(B)\text{POS}_{A-\{a\}} \neq (B)\}$$

A.5   SUPPORT VECTOR MACHINES

We will now have an overview of support vector machines (SVM): a classification approach that is used for linear as well as nonlinear data. The classification is done by constructing an n-dimensional hyperplane. The hyperplane divides the data into two classes. This hyperplane can be said to be a "boundary" or more precisely a "decision boundary" that separates the objects of one class from an other. An optimal hyperplane is selected from a set of hyperplanes that are generated. The hyperplane is found with the margins and the support vectors. Support vectors are nothing but the training sets. Kernel functions are being used by SVM, which are the class of algorithms for pattern analysis.

Figure A.1 shows that multiple hyperplanes can be drawn but the hyperplane z will be the optimal one as it maximizes the margins between the classes.

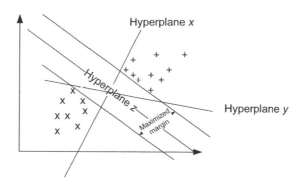

Figure A.1   Optimal hyperplane selection.

REFERENCES

1. Pawlak 1982.
2. Pawlak 1991.

Markov Processes

B.1 MARKOV PROCESSES

Definition of Markov Processes

Suppose that we perform, one after the other, a sequence of experiments that have the same set of outcomes. If the probabilities of the various outcomes of the current experiments depend (at most) on the outcome of the preceding experiment, then we call the sequence a Markov process.

A *Markov process* $\{X_t, t \in T\}$ is a stochastic process with the property that, given the value of X_t, the values of X_s for $s > t$ are not influenced by the values of X_u for $u < t$. In other words, the probability of any particular future behavior of the process, when its current state is known exactly, is not altered by additional knowledge concerning its past behavior.

A *discrete time Markov chain* is a Markov process whose *state space* is a finite or countable set and whose time (or *stage*) index set is $T = (0, 1, 2, \ldots)$. In formal terms, the Markov property is that

$$P\{X_{n+1} = j | X_0 = i_0, \ldots, X_{n-1} = i_{n-1}, X_n = i\}$$

$$P\{X_{n+1} = j | X_n = i\}$$

for all time points n and all states $i_0, \ldots, i_{n-1}, i, j$.

A particular utility stock is very stable and, in the short run, the probability that it increases or decreases in price depends only on the result of the preceding day's trading. The price of the stock is observed at 5 P.M. every day and recorded as decreased, increased, or unchanged. This sequence of observations forms a Markov process.

The experiments of a Markov process are performed at regular time intervals and have the same set of outcomes. These outcomes are called states, and the outcome of

Reinforcement and Systemic Machine Learning for Decision Making, First Edition. Parag Kulkarni.
© 2012 by the Institute of Electrical and Electronics Engineers, Inc.
Published 2012 by John Wiley & Sons, Inc.

the current experiment is referred to as the current state of the process. The state are represented as column matrices.

B.1.1   Example

Consider the following problem: Company XYZ, the manufacturer of a breakfast cereal, currently has some 25% of the market. Data from the previous year indicate that 88% of XYZ's customers remained loyal that year, but 12% switched to the competition. In addition, 85% of the competition's customers remained loyal to the competition but 15% of the competition's customers switched to XYZ. Assuming that these trends continue, determine XYZ's share of the market:

- in 2 years
- in the long run

This problem is an example of a *brand-switching* problem that often arises in the sale of consumer goods.

In order to solve this problem, we make use of *Markov chains* or *Markov processes* (which are a special type of *stochastic process*). The procedure is given below.

B.1.2   Solution Procedure

Observe that, each year, a customer can be buying either XYZ's cereal or the competition's. Hence we can construct a diagram as shown below where the two circles represent the two *states* a customer can be in and the arcs represent the probability that a customer makes a *transition* each year between states. Note the circular arcs indicating a "transition" from one state to the same state. This diagram is known as the *state-transition diagram* (and note that all the arcs in that diagram are directed arcs) (Figure B.1).

Given that diagram, we can construct the *transition matrix* (usually denoted by the symbol P) which tells us the probability of making a transition from one state to another state. Let

- state 1 = customer buying XYZ's cereal and
- state 2 = customer buying competition's cereal.

We have the transition matrix P for this problem given by

$$
\begin{array}{ccc}
\text{To state} & 1 & 2 \\
\text{From state} \quad 1 & |\,0.88 & 0.12\,| \\
2 & |\,0.15 & 0.85\,|
\end{array}
$$

Note here that the sum of the elements in each row of the transition matrix is one. Note also that the transition matrix is such that the rows are "From" and the columns are "To" in terms of the state transitions.

Now we know that currently XYZ has some 25% of the market. Hence we have the row matrix representing the initial state of the system given by

$$
\begin{array}{cc}
\text{State} \\
1 \qquad 2 \\
[0.25, \quad 0.75]
\end{array}
$$

We usually denote this row matrix by s_1, indicating the state of the system in the first period (years in this particular example). Now Markov theory tells us that, in period (year) t, the state of the system is given by the row matrix s_t where

$$
s_t = s_{t-1}(P) = s_{t-2}(P)(P) = \cdots = s_1(P)^{t-1}.
$$

We have to be careful here because we are doing matrix multiplication and the order of calculation is important (i.e., $s_{t-1}(P)$ is not equal to $(P)s_{t-1}$ in general). To find s_t we could attempt to raise P to the power $t-1$ directly, but, in practice, it is far easier to calculate the state of the system in each successive year $1,2,3,\ldots,t$. We already know the state of the system in year 1 (s_1) so the state of the system in year two (s_2) is given by

$$
\begin{aligned}
s_2 &= s_1 P \\
&= [0.25, 0.75]|0.88\ 0.12| \\
&\qquad\qquad |0.15\ 0.85| \\
&= [(0.25)(0.88) + (0.75)(0.15), (0.25)(0.12) + (0.75)(0.85)] \\
&= [0.3325, 0.6675]
\end{aligned}
$$

Note that this result makes intuitive sense. For example, of the 25% currently buying XYZ's cereal, 88% continue to do so; while of the 75% buying the competitor's cereal, 15% change to buy XYZ's cereal—giving a (fractional) total of $(0.25)(0.88) + (0.75)(0.15) = 0.3325$ buying XYZ's cereal.

Hence in year two, 33.25% of the people are in state 1—that is, buying XYZ's cereal. Note here that, as a numerical check, the elements of s_t should always sum to one.

In year three, the state of the system is given by

$$
\begin{aligned}
s_3 &= s_2 P \\
&= [0.3325, 0.6675]|0.88\ 0.12| \\
&\qquad\qquad\quad |0.15\ 0.85| \\
&= [0.392725, 0.607275]
\end{aligned}
$$

Hence in year three, 39.27% of the people are buying XYZ's cereal.

B.1.3 Long Run

Recall the question asked for XYZ's share of the market in the long run. This implies that we need to calculate s_t as t becomes very large (approaches infinity).

The idea of the long run is based on the *assumption* that, eventually, the system reaches "equilibrium" (often referred to as the "steady state") in the sense that $s_t = s_{t-1}$. This is not to say that transitions between states do not take place, they do, but they "balance out" so that the number in each state remains the same.

There are two basic approaches to calculating the steadystate:

- Computational—find the steady state by calculating s_t for $t = 1, 2, 3, \ldots$ and stop when s_{t-1} and s_t are approximately the same. This is obviously very easy for a computer and is the approach used by the package.
- Algebraic—to avoid the lengthy arithmetic calculations needed to calculate s_t for $t = 1, 2, 3, \ldots$ we have an algebraic short-cut that can be used. Recall that in the steadystate $s_t = s_{t-1}$ ($= [x_1, x_2]$, say, for the example considered above). Then as $s_t = s_{t-1}P$ we have that

$$[x_1, x_2] = [x_1, x_2] \begin{vmatrix} 0.88 & 0.12 \\ 0.15 & 0.85 \end{vmatrix}$$

(and note also that $x_1 + x_2 = 1$). Hence we have three equations that we can solve.

Note here that we have used the word assumption above. This is because not all systems reach an equilibrium, for example, the system with transition matrix

$$\begin{vmatrix} 0 & 1 \\ 1 & 0 \end{vmatrix}$$

will never reach a steady state.

Adopting the algebraic approach above for the XYZ's cereal example, we have the three equations

$$x_1 = 0.88x_1 + 0.15x_2$$
$$x_2 = 0.12x_1 + 0.85x_2$$
$$x_1 + x_2 = 1$$

and rearranging the first two equations, we get

$$0.12x_1 - 0.15x_2 = 0 \quad 0.12x_1 - 0.15x_2 = 0 \quad x_1 + x_2 = 1$$

Note here that the equation $x_1 + x_2 = 1$ is essential. Without it we could not obtain a unique solution for x_1 and x_2. Solving, we get $x_1 = 0.5556$ and $x_2 = 0.4444$.

Hence, in the long-run, XYZ's market share will be 55.56%.

B.1.4   Markov Processes Example

An admissions tutor is analyzing applications from potential students for a particular undergraduate course at Imperial College (IC). She regards each potential student as being in one of four possible states:

- State 1: has not applied to IC.
- State 2: has applied to IC, but an accept/reject decision has not yet been made.

- State 3: has applied to IC and has been rejected.
- State 4: has applied to IC and has been accepted (been made an offer of a place).

At the start of the year (month 1 in the admissions year) all potential students are in state 1.

Her review of admissions statistics for recent years has identified the following transition matrix for the probability of moving between states each month:

To		1	2	3	4	
From	1	0.97	0.03	0	0	
	2	0	0.10	0.15	0.75	
	3	0	0	1	0	
	4	0	0	0	1	

- What percentage of potential students will have been accepted after 3 months have elapsed?
- Is it possible to work out a meaningful long-run system state or not (and why)?

The admissions tutor has control over the elements in one row of the above transition matrix, namely row 2.

The elements in this row reflect the following:

- transition from 2 to 2: the speed with which applications are processed each month
- transition from 2 to 3: the proportion of applicants who are rejected each month
- transition from 2 to 4: the proportion of applicants who are accepted each month.

To be more specific, at the start of each month, the admissions tutor has to decide the proportion of applicants who should be accepted that month. However, she is constrained by a policy decision that, at the end of each month, the total number of rejections should never be more than one-third of the total number of offers, nor should it ever be less than 20% of the total number of offers.

Further analysis reveals that applicants who wait longer than 2 months between applying to IC and receiving a decision (reject or accept) almost never choose to come to IC, even if they get an offer of a place.

Formulate the problem that the admissions tutor faces each month as a linear program. Comment on any assumptions you have made in so doing.

Solution:
We have the initial system state s_1 given by $s_1 = [1, 0, 0, 0]$ and the transition matrix P given by

$$P = \begin{vmatrix} 0.97 & 0.03 & 0 & 0 \\ 0 & 0.10 & 0.15 & 0.75 \\ 0 & 0 & 1 & 0 \\ 0 & 0 & 0 & 1 \end{vmatrix}$$

Hence after 1 month has elapsed, the state of the system $s_2 = s_1 P = [0.97, 0.03, 0, 0]$.

After 2 months have elapsed, the state of the system $= s_3 = s_2 P = [0.9409, 0.0321, 0.0045, 0.0225]$

After 3 months have elapsed, the state of the system $= s_4 = s_3 P = [0.912673, 0.031437, 0.009315, 0.046575]$ and note here that the elements of s_2, s_3, and s_4 add to one (as required).

Hence 4.6575% of potential students will have been accepted after 3 months have elapsed.

It is not possible to work out a meaningful long-run system state because the admissions year is only (at most) 12 months long. In reality, the admissions year is probably shorter than 12 months.

With regard to the linear program, we must distinguish within state 2 (those who have applied to IC but an accept/reject decision has not yet been made) how long an applicant has been waiting.

Hence expand state 2 to the following states:

- 2a—a new application received
- 2b—a new application received 1 month ago.
- In this way, we never leave a new application waiting longer than 2 months— applicants in this category almost never come to IC anyway.

Hence we have the new transition matrix

$$P = \begin{array}{c} \\ 1 \\ 2a \\ 2b \\ 3 \\ 4 \end{array} \begin{array}{ccccc} 1 & 2a & 2b & 3 & 4 \\ \begin{vmatrix} 0.97 & 0.03 & 0 & 0 & 0 \\ 0 & 0 & 1-X-Y & X & Y \\ 0 & 0 & 0 & 1-y & y \\ 0 & 0 & 0 & 1 & 0 \\ 0 & 0 & 0 & 0 & 1 \end{vmatrix} \end{array}$$

Here X is the reject probability each month for a newly received application and Y the acceptance probability each month for a newly received application (these are decision variables for the admissions tutor), where $X \geq 0$ and $Y \geq 0$.

In a similar fashion, y is the acceptance probability each month for an application that was received 1 month ago (again a decision variable for the admissions tutor).

Each month then, at the start of the month, we have a known proportion in each of the states 1, 2a, 2b, 3, and 4.

Hence the equation for the (unknown) proportions $[z_1, z_{2a}, z_{2b}, z_3, z_4]$ at the end of each month is given by:

- $[z_1, z_{2a}, z_{2b}, z_3, z_4] = $ [known proportions at start of month]P; where P is the transition matrix given above involving the variables X, Y, and y. If we were to write this matrix equation out in full, we would have five linear equalities. In addition, we must have that:
- $z_1 + z_{2a} + z_{2b} + z_3 + z_4 = 1$
- $z_1, z_{2a}, z_{2b}, z_3, z_4 \geq 0$ and the policy conditions are:
- $z_3 \leq z_4/3$
- $z_3 \geq 0.2z_4$

Hence we have a set of linear constraints in the variables $[X, Y, y, z_1, z_{2a}, z_{2b}, z_3, z_4]$.

An appropriate objective function might be to maximize the sum of the acceptance probabilities $(Y + y)$, but other objectives could be suggested for this system.

Hence we have an LP that can be solved to decide X, Y, and y each month. Comments are:

- Row 1 of the transition matrix is constant throughout the year.
- This does not take into account any information we might have on how applicants respond to the offers made to them.

B.2 SEMI-MARKOV PROCESS

A *semi-Markov process* is one that changes states in accordance with a Markov chain but takes a random amount of time between changes. More specifically, consider a stochastic process with states 0, 1,..., which is such that, whenever it enters state i, $i \geq 0$: (i) The next state it will enter is state j with probability Pij, i, $j \geq$. (ii) Given that the next state to be entered is state j, the time until the transition from i to j occurs has distribution Fij. If we let $Z(t)$ denote the state at time t, then $\{Z(t), t \geq 0\}$ is called a semi-Markov process. Thus a semi-Markov process does not possess the Markovian property that given the present state the future is independent of the past. In predicting the future, would we want to know not only the present state, but also the length of time that has been spent in that state.

A Markov chain is a semi-Markov process in which

$$F_{ij}(t) = 0 \; t < 1$$
$$= 1 \; t \geq 1$$

That is, all transition times of a Markov chain are identically 1.

Let H_i denote the distribution of time that the semi-Markov process spends in state i before making a transition. That is, by conditioning on the next state, we see

$$H_i(t) = \sum P_{ij}F_{ij}(t)$$

and let μi denote its mean. That is,

$$\mu i = \int_0^\infty x\,dH_i(x)$$

If we let Xn denote the nth state visited, then $\{Xn, n \geq 0\}$ is a Markov chain with transition probabilities Pij. It is called the *embedded* Markov chain of the semi-Markov process. We say that the semi-Markov process is *irreducible* if the embedded Markov chain is irreducible as well.

Let Tii denote the time between successive transitions into state i and let $\mu ii = E[Tii]$. By using the theory of alternating renewal processes, we could derive an expression for the limiting probabilities of a semi-Markov process.

B.2.1 Proposition

If the semi-Markov process is irreducible and if Tii has a nonlattice distribution with finite mean, then

$$P_i = \lim_{t \to \infty} P\{Z(t) = i \mid Z(0) = j\}$$

exists and is independent of the initial state. Furthermore,

$$P_i = \frac{\mu_i}{\mu_{ii}}$$

B.2.2 Proof

Say that a cycle begins whenever the process enters state i, and say that the process is "on" when in state i and "off" when not in i. Thus we have an alternating renewal process (delayed when $Z(0) \neq i$) whose "on" time has distribution H_i and whose cycle time is T_{ii}.

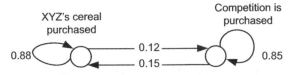

Figure B.1 Transition diagram for customer states.

B.2.3   Corollary

If the semi-Markov process is irreducible and $\mu_{ii} < \infty$, then the probability is expressed as

$$\frac{\mu_i}{\mu_{ii}} = \frac{\lim_{t \to \infty} \{\text{amount of time in } i \text{ during } [0, t]\}}{t}$$

That is, μ_i / μ_{ii} equals the long-run proportion of time in state i.

Reinforcement and Systemic Machine Learning for Decision Making, First Edition. Parag Kulkarni.
© 2012 by the Institute of Electrical and Electronics Engineers, Inc.
Published 2012 by John Wiley & Sons, Inc.

IEEE PRESS SERIES ON SYSTEMS SCIENCE AND ENGINEERING

Editor:
MengChu Zhou, *New Jersey Institute of Technology and Tongji University*

Co-Editors:
Han-Xiong Li, *City University of Hong-Kong*
Margot Weijnen, *Delft University of Technology*

The focus of this series is to introduce the advances in theory and applications of systems science and engineering to industrial practitioners, researchers, and students. This series seeks to foster system-of-systems multidisciplinary theory and tools to satisfy the needs of the industrial and academic areas to model, analyze, design, optimize and operate increasingly complex man-made systems ranging from control systems, computer systems, discrete event systems, information systems, networked systems, production systems, robotic systems, service systems, and transportation systems to Internet, sensor networks, smart grid, social network, sustainable infrastructure, and systems biology.

1. *Reinforcement and Systemic Machine Learning for Decision Making*
   Parag Kulkarni
2. *Remote Sensing and Actuation Using Unmanned Vehicles*
   Haiyang Chao, YangQuan Chen
3. *Hybrid Control and Motion Planning of Dynamical Legged Locomotion*
   Nasser Sadati, Guy A. Dumont, Kaveh Akbari Hamed, and William A. Gruver

Printed and bound by CPI Group (UK) Ltd, Croydon, CR0 4YY

27/10/2024

14580343-0001